M000222138

DATA SCIENCE

CREATE TEAMS THAT ASK THE RIGHT QUESTIONS AND DELIVER REAL VALUE

Doug Rose

Data Science: Create Teams That Ask the Right Questions and Deliver Real Value

Doug Rose
Atlanta, Georgia
USA

ISBN-13 (pbk): 978-1-4842-2252-2   ISBN-13 (electronic): 978-1-4842-2253-9
DOI 10.1007/978-1-4842-2253-9

Library of Congress Control Number: 2016959479

Copyright © 2016 by Doug Rose

This work is subject to copyright. All rights are reserved by the Publisher, whether the whole or part of the material is concerned, specifically the rights of translation, reprinting, reuse of illustrations, recitation, broadcasting, reproduction on microfilms or in any other physical way, and transmission or information storage and retrieval, electronic adaptation, computer software, or by similar or dissimilar methodology now known or hereafter developed.

Trademarked names, logos, and images may appear in this book. Rather than use a trademark symbol with every occurrence of a trademarked name, logo, or image we use the names, logos, and images only in an editorial fashion and to the benefit of the trademark owner, with no intention of infringement of the trademark.

The use in this publication of trade names, trademarks, service marks, and similar terms, even if they are not identified as such, is not to be taken as an expression of opinion as to whether or not they are subject to proprietary rights.

While the advice and information in this book are believed to be true and accurate at the date of publication, neither the authors nor the editors nor the publisher can accept any legal responsibility for any errors or omissions that may be made. The publisher makes no warranty, express or implied, with respect to the material contained herein.

Managing Director: Welmoed Spahr
Acquisitions Editor: Robert Hutchinson
Developmental Editor: Laura Berendson
Editorial Board: Steve Anglin, Pramila Balen, Laura Berendson, Aaron Black,
    Louise Corrigan, Jonathan Gennick, Robert Hutchinson, Celestin Suresh John,
    Nikhil Karkal, James Markham, Susan McDermott, Matthew Moodie, Natalie Pao,
    Gwenan Spearing
Coordinating Editor: Rita Fernando
Copy Editor: Lauren Marten Parker
Compositor: SPi Global
Indexer: SPi Global
Cover Designer: eStudioCalamar

Distributed to the book trade worldwide by Springer Science+Business Media New York, 233 Spring Street, 6th Floor, New York, NY 10013. Phone 1-800-SPRINGER, fax (201) 348-4505, e-mail orders-ny@springer-sbm.com, or visit www.springeronline.com. Apress Media, LLC is a California LLC and the sole member (owner) is Springer Science + Business Media Finance Inc (SSBM Finance Inc). SSBM Finance Inc is a **Delaware** corporation.

For information on translations, please e-mail rights@apress.com, or visit www.apress.com.

Apress and friends of ED books may be purchased in bulk for academic, corporate, or promotional use. eBook versions and licenses are also available for most titles. For more information, reference our Special Bulk Sales–eBook Licensing web page at www.apress.com/bulk-sales.

Any source code or other supplementary materials referenced by the author in this text is available to readers at www.apress.com. For detailed information about how to locate your book's source code, go to www.apress.com/source-code/.

Apress Business: The Unbiased Source of Business Information

Apress business books provide essential information and practical advice, each written for practitioners by recognized experts. Busy managers and professionals in all areas of the business world—and at all levels of technical sophistication—look to our books for the actionable ideas and tools they need to solve problems, update and enhance their professional skills, make their work lives easier, and capitalize on opportunity.

Whatever the topic on the business spectrum—entrepreneurship, finance, sales, marketing, management, regulation, information technology, among others—Apress has been praised for providing the objective information and unbiased advice you need to excel in your daily work life. Our authors have no axes to grind; they understand they have one job only—to deliver up-to-date, accurate information simply, concisely, and with deep insight that addresses the real needs of our readers.

It is increasingly hard to find information—whether in the news media, on the Internet, and now all too often in books—that is even-handed and has your best interests at heart. We therefore hope that you enjoy this book, which has been carefully crafted to meet our standards of quality and unbiased coverage.

We are always interested in your feedback or ideas for new titles. Perhaps you'd even like to write a book yourself. Whatever the case, reach out to us at editorial@apress.com and an editor will respond swiftly. Incidentally, at the back of this book, you will find a list of useful related titles. Please visit us at www.apress.com to sign up for newsletters and discounts on future purchases.

—*The Apress Business Team*

For Jelena and Leo

Contents

Part V: Storytelling with Data Science 189

Part VI: Finishing Up 231

About the Author

Doug Rose specializes in organizational coaching, training, and change management. He has worked over twenty years transforming organizations with technology, training and helping large companies optimize their business processes to improve productivity and delivery. He teaches business, management, and organizational development courses at the University of Chicago, Syracuse University, and the University of Virginia. He also delivers courses through LinkedIn Learning. He is the author of *Leading Agile Teams* (PMI Press, 2015) and has an MS in Information Management and a JD from Syracuse University, and a BA from the University of Wisconsin-Madison. You can follow him at https://www.linkedin.com/in/dougrose.

Acknowledgments

First and foremost, I'd like to thank my wonderful wife and son. My wife is still my top proofreader. Her love and support drives me to be better. My son's love of writing is my inspiration. He's currently finishing his sequel to the well-received *Joe* series, *The Adventures of Joe Part 2: The Death of John* (2016, publisher forthcoming). I'd also like to thank my literary agent Carole Jelen for her great work and unwavering professionalism. I'd like to thank all the wonderful people at Apress publishing. This includes editor Robert Hutchinson, coordinating editor Rita Fernando Kim, and developmental editor Laura Berendson. Much of this work is based on previous courses I've taught at the University of Chicago, Syracuse University, University of Virginia, and LinkedIn Learning. At LinkedIn I'd like to thank content manager Steve Weiss and senior content producer Dennis Meyer, along with content producer Yash Patel and directors Tony Cruz and Scott Erickson. At the University of Chicago, I'd like to thank Katherine Locke, and at Syracuse University, special thanks to my graduate students along with Angela Usha Ramnarine-Rieks and Gary Krudys. I received some terrific help and guidance on this book from editor Mary Lemons. I also want to thank the great Lulu Cheng for help with the data visualizations and reports. Finally, I want to give a special thanks to all the wonderful companies that I've worked for over the years. Many of the ideas for this book came from the feedback that I've received while working as a management coach. I owe a special thanks to The Home Depot, Cox Automotive, Paychex, Cardlytics, Genentech, and The United States Air Force Civil Air Patrol, along with federal and state government agencies in both Georgia and Florida.

Introduction

After college, one of my first jobs was working for Northwestern University's Academic Computing and Network Services (ACNS). It was 1992, and the lab was an interesting mix of the newest technology. I remember the first time we tried the World Wide Web (WWW) on Steve Job's NeXTcube. It was just a year after the first web servers were available in Europe. We were underwhelmed as we watched the graphics slowly load on the small gray screen. None of us understood why anyone would wait to see an image. You could instantly find what you were looking for with text browsers like TurboGopher. Why would anyone wait ten seconds for a button that says "Click here?"

Despite our dire predictions, the World Wide Web took off. Students poured in and asked for demonstrations. We were given coveted webspace for personal HyperText Markup Language (HTML) pages. My page was simple. It was a small scanned image with my new e-mail address. I used the name of the messenger god: hermes@merle.acns.nwu.edu. At the time, there couldn't have been more than a few hundred pages like it on the web.

After a few years, I "dreamed away the time" and learned skills that I thought were only useful in academia. We were caught off guard when a few business recruiters called in and asked our staff what we knew about the web. They wanted to know if we were "HTML programmers." A few of us shrugged and listened to the list of requirements. Did we know the World Wide Web? Did we know how to create pages in HTML? Did we know how to network computers using TCP/IP? Each one of us said, "Yes, yes, and yes." Before we knew it, most of us were whisked into Chicago skyscrapers. Our titles changed to "web developers" and we traded in our shorts and T-shirts for oxfords and chinos.

My first "developer" job was for Spiegel, a large women's clothing catalog. I helped train copywriters on how to use HTML to create their first e-commerce site. I remember telling the copywriters that soon everyone would learn how to create HTML pages. That instead of QuarkXPress, we would all be churning out HTML. The road to their web-connected future was paved with HTML. They needed to give up their rudimentary tools and understand high-tech alternatives such as Microsoft's FrontPage.

I warned them that in order to stay relevant, they needed to learn new tools and software. I explained the benefits of hand coding HTML. They needed to learn how to create an HTML table from scratch. They patiently watched as I showed them how to type in <table>, <tr>, and <td>. My reasoning for teaching them this was pretty simple. You need to go deep into the tools to get the benefits of the technology. Copywriters, graphic designers, trainers, and managers would all need to know the basics of HTML.

But it didn't turn out that way. We didn't all become HTML programmers. In fact, most people today wouldn't recognize an HTML page. Yet we fully participate in the vision behind the World Wide Web. Our managers, graphic designers, and even grandparents are sharing information in ways that could've never happened using simple HTML. In a sense, none of us became HTML programmers, and yet we all became web developers. We didn't learn more about the tools; instead, we learned more about the value of the web. It became possible to share information in real time. With a click of the button, you could publish your thoughts around the world. At the time, this concept was difficult to imagine. It was an entirely new mindset.

Still, my warning about a future filled with HTML was not a complete waste. It was just misguided. I learned that technology is transient. The software and tools are important, but it's the things you learn from these tools that actually last the longest. In a way, the tools are a vehicle to a larger mindset. Instead of focusing on the tools and technology, I should've helped the copywriters shift their mindset. What does it mean to share information in real time? What will be the challenges and opportunities with this new technology? The ones who did pick up on this were able to create some of the first blogs, e-commerce, and online catalogs.

Fast-forward to today. It's been over a quarter century, and a new generation is being whisked into skyscrapers. The data science recruiters are also pulling from academia. These young biologists, statisticians, and mathematicians are getting their own phone calls. "Do you know data science? Do you know how to use R and Python? Do you know how to create a Hadoop cluster?" They're the first round of hires in a world that needs data scientists. Once again, the focus is on the tools and software. Everyone will need to know how to use R or Python to participate in this growing field. The future is paved with complex data visualizations.

But it won't turn out that way. The future of data science won't be filled with data scientists. Instead, many more people will have their careers enhanced with data science tools. The data scientist of the future will be today's graphic designers, copywriters, or managers. The data science tools will become as easy to use as the web publishing tools you use today. The data science equivalents of web tools like Facebook, LinkedIn, and WordPress are probably just a few years away.

The most lasting thing you can do today is change your mindset and embrace the *value* in data science. It's about enhancing our understanding of one another. The technology allows you to gain insights from massive amounts of data in real time. You'll be able to see people's behavior at an individual and group level. This will create a new generation of tools that will help understand people's motivations and communicate with them in more meaningful ways. So what does it mean to be able to crunch this kind of data in real time? The first one to understand this will create some of the top data science trends of the future.

That's why this book takes a different approach to data science. Instead of focusing on tools and software, this book is about enhancing the way you think about this new technology. It's about embracing a data science mindset. That's how you can get long-term value. You can start applying data science ideas to your organization. Becoming an expert in R, Python, or Hadoop is terrific. Just keep in mind that these tools are best if you're interested in being a statistician, analyst, or engineer. If you're not interested in these fields, it might not be the best use of your time. You don't have to know how to mix concrete to be an architect. The same is true with data science. You could work the business side of the team without having to know statistical software. In fact, in the future, many more people from business, marketing, and creative fields will participate in data science. These teams of people will need to think about their work in a different way. What kind of data might be valuable? What type of questions will help your organization? These are the skills that will have lasting value well beyond any one toolset.

That's why you should think of this book as having three overarching concepts:

- The first is that you should *mine* your own company for talent. You can't change your organization by hiring data science heroes. The best way to get value from data science is by changing part of your organization's focus from managing objectives to researching and exploring.

- The second is that you should form small agile-like data teams that focus on delivering insights early and often.

- Finally, you can only make real changes to your organization by telling compelling data stories. These stories are the best way to communicate your insights about your customers, challenges, and industry.

Much of the "science" in data science comes from the scientific method. You're applying a scientific method to your data. This is an empirical approach to gaining new knowledge and insights. An empirical approach is where you gain new knowledge from observation and experimentation. When you dip your toe in the pool, you are using an empirical approach. You're running a small experiment and then reacting to the results. If the water's too cold, you work on your tan. If the water's warm, you can jump right in.

You don't have to be a statistician to be able to ask interesting questions or to run a small experiment. Many people in different fields can contribute to this method of inquiry. In fact, you often get the best questions and feedback when you have people from diverse backgrounds.

This book divides the three big concepts into five parts. Each part is a skillset that you'll need for a data science mindset.

- Part I goes into the language and technology behind data science.

- Part II is about building your data science team.

- Part III is about how your team will work together to deliver insights and knowledge.

- Part IV is about how a data science team should *think* about data.

- Part V helps you tell an interesting story. Most scientists will tell you that your results won't mean much if you can't communicate your story.

Part I is foundation material that will help you work in this field. It's not meant to turn anyone into a statistician or data analyst. Instead, you get a basic overview of some of the key concepts in data science. This is an important first step. If you think about the web example, even with modern tools, you need to have an understanding of the key concepts to contribute to the web. You need to know what it means to upload. You also need to know basic file formats like GIF and JPEG. These might seem like common terms, but they weren't when the web first started. Part I is about understanding data science key terms and being able to communicate with the data analysts in your organization.

Part II is about building your data science team. Many organizations believe that they should hire superheroes to help them get to the next level in data science. The pool of data science stars is small, and because of this, many people are trying to "skill up" to become a hero on their own. The strategy might work in the short term, but a lot of data suggests that these heroes cause more harm than good. There's strong evidence that suggests that an organization gets a lot more value from building up existing talent.[1] In this

---

[1] Boris Groysberg, Ashish Nanda, and Nitin Nohria, "The Risky Business of Hiring Stars." *Harvard Business Review* 82, no. 5 (2004): p. 92-101.

part, you learn about the different roles that you'll want to create for data science teams and some common practices on how these team members can work together.

Part III goes into how your team will deliver valuable knowledge and insights. Many data science teams are just starting out, and they're still in a honeymoon period. They can work in the twilight areas of your organization. Most companies are waiting to understand the team before they scrutinize the work. It won't take long for key business people in your organization to start questioning whether or not your team is delivering business value. There is already evidence that many teams are still ignoring the simple strategy for self-preservation.[2]

You also see a simple process for how to deliver predictable value. Data science mirrors some of the challenges you run into when developing complex software. Your team can benefit from delivering value frequently and making quick pivots when you learn something new. So this part goes through how to deliver data science insights in "sprints." These are quick, iterative, and incremental bits of data science value improved and delivered every two weeks.

This book is geared towards data science teams. The focus is on giving the team a shared understanding of data science and how they'll work together to deliver key insights. In the paper *The Increasing Dominance of Teams in Production of Knowledge,* professors from the University of Miami and Northwestern University showed that there is a strong trend toward teams as the primary way to increase organizational knowledge.[3] In the last five decades, teams of people have created more patents and frequently cited research than individual inventors and solo scientists working in a lab. The trend in scientific research has been away from working with heroes. Some of the best work is coming from teams of 3-4 people. The same is true with data science. You can get better insights from small groups over one or two heroes.

This book gives you a broad survey of many of these topics, but it isn't intended to be a deep dive into any one of them. Instead, you'll see a strategy for bringing them together to deliver real value. There are already plenty of resources out there on specific practices. If you're a data analyst, there are books on R, Python, and Hadoop. There are also extensive resources on data visualization and displaying quantitative information. There are footnotes if you want to learn more on any topic.

---

[2]Ted Friedman and Kurt Schlegel, "Data and Analytics Leadership: Empowering People with Trusted Data," in Gartner Business Intelligence, Analytics & Information Management Summit (Sydney, Australia: Gartner Research, 2016).
[3]Stefan Wuchty, Benjamin F. Jones, and Brian Uzzi. "The Increasing Dominance of Teams in Production of Knowledge." *Science* 316, no. 5827 (2007): p. 1036-1039.

You'll also see a lot of data visualizations in this book. Each of these includes a link to the source code. The links are shortened using the URL `http://ds.tips` along with a five-character string. That way, it's easier if you don't have the ability to copy/paste. Again, the point of these visualizations is not to teach you how to use these tools. Instead, it's to give you a starting point if you want to build on any of the included visualizations. The main purpose of having these reports is to give you a sense of what it means to be on a data science team. These are the types of charts and reports you should expect from a data analyst. You can see typical charts that will help you understand the data. You will also get a sense of the different types of questions you can ask. I tried to use different toolsets for many of the visualizations. Some of them use the programming language R and others use Python, with some of the add-on libraries. There are also a few outside web sites that can help you create helpful word clouds and maps.

Part IV goes into a key component of the scientific method. You'll have to think about your data using key critical thinking techniques. The data will only show you things that you're prepared to see. Critical thinking and reasoning skills can help you expand the team's ability to accept the unexpected. There are plenty of examples of individuals, teams, and organizations looking at data and seeing what they expect without questioning their reasoning. This type of thinking leads to many false conclusions. The field of data science is poised to make this problem even worse. Bad reasoning can create a false foundation that will weaken all of your future insights.

The creative engine behind critical thinking is asking the right questions. Part IV also goes into different types of questions and how each type can help you find insights. There are the broader essential questions that can help you tackle larger concepts. Then there are nonessential questions that help you build up knowledge over time. You'll also see the best way to ask these questions. When your team works together, they often assume that someone else will answer an essential question. You'll see strategies for working together as a team to root out assumptions and find new areas to explore.

You'll see the value in taking the empirical approach to exploring your data. This approach works well with data. In fact, the volume of data is so great and changes so often that you're often forced to use an empirical approach. Instead of making a few grand theories, you're forced to stumble into your answers by asking dozens or even hundreds of small questions and running dozens of experiments.

Part V is about data storytelling. This is something that doesn't always come easy to data science teams. Data analysts, business managers, and software developers don't usually have the best background for creating a compelling story. Yet telling stories is one of the best ways to communicate complex information. Often, good science will suffer because it isn't *told* well to an outside audience. The challenge for your data science team is to take their reasoning, insights, and analysis and roll it all up into a short, simple narrative. In data science, you're often reconstructing the behavior of thousands or even millions of individuals. Their behaviors are not always driven by rational actions. Charts and analysis can show what people do, but it can't always show *why* they do it. In most cases, the *why* is much more valuable when you are trying to gain business insights.

This part is a high-level overview of what it takes to create a compelling story. You'll see how to weave together a plot, conflict, and resolution to re-humanize the data and reconstruct your customers' motivations. Most teams place too much emphasis on creating beautiful charts and graphs. They figure if the data is well designed, the story will tell itself. That's why there's so much material available on how to create elegant data visualizations. The reality is that few people remember the charts and graphs. People are more likely to remember the stories you tell.

These five parts together should help your team think about data in a way that will bring more value to your organization. The new tools and software will allow your teams to explore new areas in a way that, until recently, was technically impractical. Still, these tools are not going to provide much if your team can't think about the data in a new way. In the past, the technology limited your team's creativity. Now, you'll have the ability to ask new questions. What does my customer really want? What's the real value of my brand? What new product will be a success? It's the creativity of your questions and the stories you tell about your insights that will help you extract the most value from your data.

What questions will your teams ask?

Defining Data Science

In this part, we'll put four walls around the term "data science." You won't have any trouble finding people to agree that data science is important. It's much harder to find more than a few people who have a shared definition of data science. We'll start by introducing databases and how data science uses different data types. Then you'll see how to apply statistical analysis to these different types of data.

Understanding Data Science

In this chapter, I begin by defining what a data scientist is and what he or she does. Then I cover the different types of software and tools used to collect, scrub, and analyze data. Once you know the different types of software and tools used in data science, I'll briefly go over the importance of focusing on organizational knowledge.

Defining a Multi-Disciplinary Practice with Multiple Meanings

So what is a data scientist? A data scientist is a little more difficult to define than other types of scientists. If you're a political scientist or a climate scientist, you have a degree from an established program. The term "data scientist" became widely used before "data science" was a well-defined discipline. Even now, people who call themselves data scientists come from a variety of different fields. As a discipline, "data science" is still sorting itself out. It's a bit like early archaeology. Anyone could call themselves an archaeologist as long as they picked up a shovel and started digging for artifacts. These days, you have to go through a university and spend years doing research to become an archeologist. Like early archaeology, data science is still more of a practice than a discipline.

© Doug Rose 2016
D. Rose, *Data Science*, DOI 10.1007/978-1-4842-2253-9_1

You're a data scientist if you work with your data in a scientific way. Whether you choose to call yourself a data scientist is still pretty much up to you. There are certainly groups of people who are a better fit for the title "data scientist" than others. If you're a statistician or a data analyst, or you work in one of the biological sciences, you can probably argue that you've always been a data scientist. Some of the very first people to call themselves data scientists were actually mathematicians; others came from systems and information engineering and some even came from business and finance. If you worked with numbers and knew a little bit about data, you could easily call yourself a data scientist.

Now, with the increased demand for data scientists, there will be more movement to create a standardized skill set. You're already starting to see this with new programs at Berkeley, Syracuse University, and Columbia University. New degree programs will allow companies to rely on a common skill set when they're hiring. For now, that's not the case. In fact, there's still some danger that data scientists will be seen as anyone who works with data and can update their LinkedIn profile.

The best way to think about data science is to focus on the science and not the data. In this context, the science uses a scientific method. You should run experiments and see the results using an empirical approach. Empiricism is one of the ways that scientists gain insights and knowledge by reacting to the data through experiments and questions. A data scientist should use this skill every day. An empirical approach is a combination of knowledge and practice. You probably use the empirical approach and don't realize it.

As a coach and trainer, I have to do a fair amount of traveling. That usually means that I find myself in different hotels. I'm always amazed at how many different types of faucets and fixtures there are in the world. One thing I always struggle with is how to deal with the complexity of hotel showers. After awhile, I realized that the best way to deal with the problem is to use an empirical approach. First, I have to guess how to turn the shower on. I start by asking an empirical question. How do I turn on the shower? Then I try an experiment. When I press one button, water fills the tub. If I press another, the showerhead springs to life. After I get the water on, I have to twist and turn the different knobs to see if I can control the temperature. If I twist one knob, it gets too hot. If I twist another, it gets too cold. So I ask questions and reevaluate until I can make the water comfortable. I wouldn't want to use a theoretical approach. I could theorize on how to make the water comfortable, and then I could flip a dial and jump in the shower. The problem is that I'd be likely to be frozen or scalded.

Data scientists use this same empirical approach all the time. They ask questions of the data and make small adjustments to see if they can gain insights. They turn the knobs and ask more interesting questions.

For the purposes of this book, I focus on a data scientist as someone who uses the empirical approach to gain insights from data and focuses on the scientific method. Our emphasis is the *science* in "data science" and not the data.

Using Statistics & Software

Because data science is still defined by practice, there's an extra emphasis on using common tools and software. Remember that data scientists are like the first archaeologists. So think of the software as the brushes and pickaxes you need to make discoveries. However, try not to focus too much on learning all the tools, because they are not all you need to know. The scientific method is what makes someone a data scientist, not the tools. The tools needed by data scientists fall into three general categories:

- **Software to hold the data:** These are the spreadsheets, databases, and key/value stores. Some popular software includes Hadoop, Cassandra, and PostgreSQL.

- **Tools used to scrub the data:** *Data scrubbing*, also called *data cleansing*, makes data easier to work with by modifying or amending data or deleting duplicate, incorrectly formatted, incorrect, or incomplete data. Typical tools used to scrub data are text editors, scripting tools, and programming languages such as Python and Scala.

- **Statistical packages to help analyze the data:** The most popular are the open source software environment R, IBM SPSS predictive analytics software, and Python's programing language. Most of these include the ability to visualize the data. You'll need this to make nice charts and graphs.

Holding the Data

Let's first look at the tools you need to know to hold the data. One term that you'll hear a lot is big data. Big data sounds like the title of a 1960s horror movie. You picture some screaming woman in cat-eye glasses being swallowed up by an oozing mountain of data. **Big data** is data sets that are so large they won't fit into most data management systems. Some people confuse data science and big data because they have been hyped at the same time and often mushed together. However, remember that data science is applying the scientific method to data. This doesn't assume that your data has to be big. In fact, there's a good book called *Data Smart: Using Data Science to Transform Information into Insight* by John W. Foreman,[1] which introduces data science statistics using only spreadsheets.

---

[1] Foreman, John W. *Data Smart: Using Data Science to Transform Information into Insight.* John Wiley & Sons, 2013.

Nevertheless, one of the most active areas in data science is around big data, and there is software designed specifically to handle big data. The open-source software package **Hadoop** is currently the most popular. Hadoop uses a distributed file system to store the data on a group of servers, typically called a **Hadoop cluster**. The cluster also distributes tasks on the servers so that you can run applications on them as well. This means that you can store petabytes of data on hundreds or even thousands of servers, and run processes on the data in the cluster. The two most common processes that run on Hadoop clusters are MapReduce and Apache Spark. MapReduce works with the data in batches and Spark can process the data in real time.

Scrubbing the Data

After you have collected data, you most likely want to use some tools to scrub the data so it's more usable. Scrubbing the data makes it easier to work with by modifying or amending data or deleting duplicate, incorrectly formatted, incorrect, or incomplete data. Imagine that you are collecting millions of your customer's tweets, which can contain text, pictures, and even videos. When collecting this data, you could create a script that divides all of the incoming tweets into the various types (text, pictures, videos, and others). This would allow you to analyze each of these groups separately and using different parameters. If you do this analysis frequently, it may be better to create a small Python application to perform the operation on the cluster instead of a script that does it as the tweets come in.

Data scientists can spend up to 90% of their time adapting and scrubbing their data to make it more usable, so automating this process is critical to this step.

Analyzing Data

The final group of tools is the ones used to analyze the data. Two of the most popular are R and Python.

R is a statistical programming language and software environment that allows you to make connections and correlations in the data, and then present them using R's built-in data visualization. This allows you to have a nice diagram for your report. For example, imagine your company wants a report to see whether there's a connection between their positive feedback and whether that feedback occurs in the daytime or at night. One way to gather this information is to capture Twitter data in your Hadoop cluster, and then use data scrubbing to categorize the tweets as positive or negative. Next, you could use a statistical package like R to create a correlation between the positive and negative tweets and the time they were posted, and print a report showing the results in a nice diagram.

Keep in mind that these are the most popular tools. If you are part of a data science team, you almost certainly hear at least one of them come up in conversations. There are many more tools that automate collecting, scrubbing, and analyzing the data.

There are many organizations that are spending a lot of money trying to buy their way into this space. Try to remember to focus on the analysis. The data and tools are just the vehicle to gain greater insight. Spend money with caution in this growing field.

Uncovering Insights and Creating Knowledge

Over the last 20 years, most organizations have focused on increasing their operational efficiency to be leaner and more flexible by streamlining their business processes. They asked operational questions such as, "How can we work better together?"

Data science is different; it isn't objective-driven. It's exploratory and uses a scientific method. It's not about how well an organization operates; it's about gaining useful business knowledge. With data science, you ask different types of questions, such as:

- What do we know about our customer?
- How can we deliver a better product?
- Why are we better than our competitors?

These are all questions that require a higher level of organizational thinking, and most organizations aren't ready to ask these types of questions. They are driven to set milestones and create budgets. They haven't been rewarded for being skeptical or inquisitive.

Imagine that you're in a business meeting and someone asks these questions. Why are we doing it this way? What makes you think that this will work? Why is this a good idea? Chances are, the person asking this would be seen as annoying. Usually, someone will answer with something like "Didn't you read the memo?" However, these are the skills you need to build organizational knowledge. These are the questions you want from your data science team. Still, most people in organizations are focused on getting things done. Questions like those mentioned are seen as a barrier to moving forward. However, as an organization, you gain knowledge by asking interesting questions.

I once worked for a web site that connected potential car buyers to dealers. There were hundreds of information tags on the web site, which showed whether the customer was hovering or clicking on their links. All of this data flowed into a Hadoop cluster, and there were terabytes of this data every week. The company had historical data going back years. They spent large sums of money and even set up departments to focus on collecting and maintaining this data. Collecting data was easy. The software they used was simple and easy to create. The hard part was figuring out what to do with the data.

This seems like a common challenge for many organizations starting out in data science. These organizations see it mostly as an operational challenge. They focus on the technical side of the data. It's about collecting the data because it's relatively cheap and easy to understand. It's meeting-friendly and everyone can get behind the effort. They'll even create multiple clusters or data lakes to pool their data from all over the organization. That's the easy part. What organizations struggle with is the science. They're not used to asking and answering interesting questions.

Think about the experiments and questions you could ask if you were the data scientist for this car web site. You could run an experiment like in Figure 1-1 that changed the colors of the images to see whether customers were more likely to click on an image if it were red, blue, or yellow. If the reports showed that customers are 2% more likely to click on a car if it's red, the organization could share that with car dealerships to generate new revenue. You could run an experiment to see if the site has too many cars on a page by reducing the number of cars displayed. Then, you could run a report to see if the likelihood that a customer would click on the link increased with fewer cars displayed.

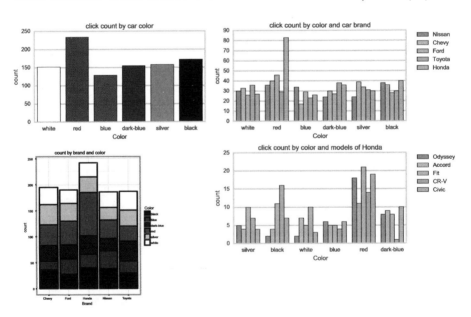

Figure 1-1. Changing the color of the cars

Note   See how to create these charts at http://ds.tips/4H5Ud.

This is the type of empirical research that a data scientist should be thinking about. They should twist the dials of the data, ask interesting questions, run quick experiments, and produce well-designed reports.

Summary

In this chapter, you learned what a data scientist is and what he or she does. You also found out about different types of software and tools used to collect, scrub, and analyze the data, as well as how to uncover insights and create knowledge. The key is to ask interesting questions. In Chapter 2, you'll learn about some database basics.

Covering Database Basics

As you have seen, data science revolves around better understanding your data. That's why you'll work with databases to access the data you need to ask interesting questions. There are many different types of databases. In addition, there is a lot of terminology used specifically for databases. You will need to be familiar with the basic concepts and terms used in the database world and with how different databases are organized.

Making Connections with Relational Databases

Data scientists will work with data in many different forms, including legacy databases or old spreadsheets. They may also work with pictures and video. As a data scientist, you should be familiar with common ways that organizations store their data. Most organizations have a wide range of databases. Some of them are very modern—others, not so much.

The best way to understand these different technologies is to start from the beginning. Even the most modern databases are often built on technology that is over 50 years old. Modern databases really got started with the Apollo space mission in the late 1960s. The rockets that would go to the moon required millions of parts, and NASA worked with IBM to create an Information Management System, or IMS, to organize this data. The space agency had early manifests, which looked very much like a modern spreadsheet. They were computer files with series of columns and long lists of rows.

© Doug Rose 2016
D. Rose, *Data Science*, DOI 10.1007/978-1-4842-2253-9_2

As you can imagine, a table with million rows might become difficult to manage on a small black and white terminal. Right around the same time, the space agency used the first relational databases. These databases divided the data into groups of tables. Each of these tables still looked like a spreadsheet, but represented a smaller chunk of the data.

Then they created relationships between those tables. Instead of having one long list of a million parts, they could create 50 tables with 20,000 parts each. That's why these are called relational databases. The database is based on groups of tables that have a relationship with each other. Even early database engineers struggled to create an efficient way to group database tables. They created maps to show how the tables related to each other. They call these maps **schemas**. A schema is what makes your relational database either easy to use or a nightmare to manage.

Even with these early databases, you can see how the engineers might've struggled with creating schemas. Should they create tables around the largest parts? Maybe make a table for just the thrusters, then another table for the fuel tank? The problem here is that if you change the design of the rocket, then you also have to change the design of the database.

Maybe you could create tables based on the manufacturer of the part. The problem with that is that you might have a manufacturer that produces thousands of parts and another manufacturer that only produces a few dozen. This continues to be a challenge today. Relational databases require a good amount of upfront design. You have to know a lot about what your data will look like before you start collecting. If you're wrong, it takes a lot of effort to redesign your database.

IBM later commercialized the IMS that they created for NASA. In the mid 1970s they developed a **Structured Query Language (SQL)** to help their customers pull data from the system. This language is still very popular. SQL is an elegant language that can pull data from several different relational tables. It reconnects all the different tables and presents the data as if it was all stored in one large sheet. This virtual table is commonly called a "**view**."

A lot of functionality has been added to relational databases over the years. They gave rise to the relational database management system (RDBMS). Companies like IBM, Microsoft, and Oracle still support and develop relational database management systems.

> ▨ **Note**   Another relational database term you may hear is **CRUD**, which stands for create, read, update, and delete. It describes all the functions of a RDBMS. Sometimes people put an S in front for "search," and use the acronym **SCRUD**.

Getting Data into Warehouses Using ETL

The terms and concepts discussed in this section are used by data science teams. Try not to get overwhelmed by the language. If you understand the terms and challenges, you're more likely to quickly get the data you need.

Many data science concepts build on previous work with relational databases. Companies have been capturing and trying to analyze data for decades. Even today, RDBMS is still the cornerstone of enterprise databases, and you need to understand RDBMS terms for data science projects. One place you're likely to run into RDBMS terms is when working with an **enterprise data warehouse (EDW)**. An EDW is a special type of relational database that focuses on analyzing data. Traditional databases are optimized for **online transaction processing (OLTP)**. An EDW is used for **online analytical processing (OLAP)**.

Think of it this way: a typical database is focused on working with data in real time, while an EDW is focused on analyzing what already happened.

Let's imagine you have a web site that sells running shoes. You hired an engineer to create your database. She created dozens of different tables and relationships. There's a table for customer addresses, a table for shoes, a table for shipping options, and so on. The web server uses SQL statements to gather the data. When a customer buys a pair of shoes, their address data is associated with the shoe, the web server provides the customer with his or her shipping options, and the pair of shoes is shipped. You want this database to be fast and efficient, and to focus on speedy returns. This database is what your customer is doing in real time. This is an OLTP database.

You also ask your database engineer to create a script that uploads the data each day to your warehouse. Your data warehouse is optimized for analytical processing. This is an OLAP database focused on creating reports.

You have a data analyst create a report to see if there's any connection between a customer's address and the types of shoes they buy, as shown in Figure 2-1. You find that people in warmer areas are more likely to buy brightly colored shoes. You use this information to change your web site so customers from warmer climates see lighter shoes at the top of the page.

Color Brightness Average by State

Figure 2-1. Color brightness average by state

▓ **Note**   See how to create this chart at `http://ds.tips/trAp8`.

Now say that your web site becomes really successful and is bought by a company that sells all types of sports clothing. That company has one warehouse for all of their web sites, and they want to combine the data from your web site with all of their other web sites. At this point, the company will do something called an **ETL**, which stands for **extract, transform, and load**. They pull the data from your web site and then load the data into their EDW.

When they extract your data, they will attempt to do so in some standard format so they can transform the data into something that works well with their data warehouse. The tables from their warehouse might have a different schema. For example, the data warehouse might have shipping information in the customer table, whereas your database has the shipping information in its own table. The data has to be transformed to get it into the EDW. A data analyst will most likely spend most of his or her time scrubbing and joining the data so that it will fit, and then, finally, load the transformed data into the warehouse.

The previous scenario isn't the only one in which you may need to perform an ETL. Some companies may have a data warehouse that is separate from the Hadoop cluster, in which case they will need to run an ETL on the warehouse data to move it into the Hadoop cluster. In this case, the data analyst needs to transform the data so it can be used in the cluster.

A lot of organizations often see Hadoop as a replacement for costly data warehouses. They want to save money by storing data on inexpensive hardware instead of a costly warehouse appliance. In this case, companies may rewrite their ETL procedures so they can load data into a Hadoop cluster, and then phase out or shut down the warehouse.

Letting Go of the Past with NoSQL

Often, a data science team needs a more flexible way to store its data. Remember that relational databases rely on a schema. You need to know a lot about your data before you can put it in the database, which means that you have to plan ahead. You have to know what type of data appears in the database fields (text, video, audio, or other), organize these fields into tables, and then create the table relationships. The database needs a set structure so you can create, read, update, and delete your records. For some extremely large databases, this overhead can bog down your servers.

Let's go back to your running shoe web site. A customer finds a pair of shoes and goes to the checkout page. At this point, the web site joins the pair of shoes being purchased with the customer's address. This checkout page needs access to four different database tables:

- The shoe table
- The customer table
- The address table
- The shipping table

That's a lot of work for a database. The harder your database works, the slower your web site.

How do you speed things up? Do you need to buy a bigger server, split your tables on several servers, or have several servers that synchronize across the network? For really large web sites, these options can seem unnatural.

Now, imagine a database that stores everything in the checkout page as one transaction. A database transaction is one all-or-nothing chunk of work that must complete. A record is created for the pair of shoes, your customer, their address, and shipping all in one shot. Now what if the data isn't split into tables and no relationship needs to be queried? The information is just dumped in and you're done.

That's the idea behind NoSQL. NoSQL was first used as a Twitter hashtag for developers who wanted to move beyond relational databases. It's actually not a slam against SQL. In fact, NoSQL doesn't have very much to do with SQL at all. It's more about the limitations of relational databases. In general, a NoSQL database should be non-relational, schema-less, cluster-friendly, and, hopefully, open source.

All of these qualities should appeal to a data science team. A database that isn't relational is typically easier to change and simpler to use. There doesn't have to be a disconnect between the way your web application looks and the way you store your data. You also won't have to go through the ugly process of creating and splitting tables that already exist. This is commonly referred to as normalizing your database. Without a schema you don't have to worry about knowing everything upfront.

Back to the running shoe web site. It was bought by a larger company. This company wants to add your customers to their frequent buyer program. With a relational database, this is a serious architectural challenge. Should you have the frequent buyer number in the customer table? Do you need to create an entirely new table to just store frequent buyer numbers? Can a customer have more than one buyer number? Can two customers share the same number? Before the customers can be added to the frequent buyer program, all of these questions need to be sorted out. You have to rework the database and figure out how to correct for missing data.

Without a schema, new fields become almost trivial. You just store it as one transaction. If the customer has a frequent buyer number, it appears in the transaction. If the customer doesn't have one, the field doesn't exist.

Finally, a NoSQL database should be cluster friendly. You should be able to store the data in several hundred or even thousand database servers. In NoSQL, the record saved in a transaction is called an **aggregate**. These aggregates hold all of the data: the shoe, customer, address, and shipping information. These aggregates are easier to synchronize across many database servers. Most of the servers work in clusters. This allows them to synchronize amongst each other and then send out updates to other clusters.

Note   The word "cluster" should sound familiar. It's the same way that Hadoop works with its data sets. In fact, much of Hadoop is built on HBase, which is an open source NoSQL database.

When you're working in a data science team, you're almost certain to run into NoSQL. For many organizations, this is the preferred way to deal with large data sets. Because of its simpler design, it's also much easier for developers to create web applications that can quickly grow into an enterprise scale.

The Big Data Problem

As mentioned earlier, big data and data science have been so intertwined that many organizations see them as one and the same. Remember that data science is using a scientific method to ask interesting questions. That doesn't mean that you need a lot of data to ask these questions. Big data provides a rich new source of data that allows you to ask questions that couldn't be answered with a smaller data set.

Big data isn't really a noun. In the original NASA paper,[1] it was described as a "big data problem." You can read this one of two ways: it's a "big data" problem or a big "data problem." If you read the whole paper, it seems that they put the emphasis on the problem. It's not about "big data." It's about the problem of what to do with all this new data. This is also addressed a decade later with the McKinsey report.[2] In the report, the authors refer to big data as data that exceeds the capability of commonly used hardware and software.

So why is it important to think of big data as a problem and not a noun? Well, it's because many companies that start big data projects don't actually have big data. It might seem like it's big because there is a lot of it. It also seems like a problem because it's a real challenge to store and collect. But it's not a "big data problem."

One way you can determine whether you have a big data problem is to see if your data falls into four categories. You can remember these categories as the four V's. Ask yourself these questions:

- Do I have a high volume of data?
- Do I have a wide variety of data?
- Is the data coming in at a high velocity?
- Does the data I'm collecting have veracity?

To be big data, it needs to have all four of these attributes.

The volume question is usually pretty easy. If you're collecting petabytes of data each day, you probably have enough volume. Of course, this might not always be a problem. In the near future, maybe an exabyte will be considered a high enough volume to be a problem.

There should be a wide variety of information. There can be text, videos, sound and pictures.

For velocity, think of the New York Stock Exchange. They handle billions of transactions each day. They have a high volume of data coming in at a high velocity. The stock prices are pouring in and fluctuating in milliseconds.

---

[1]Cox, Michael, and David Ellsworth. "Application-controlled demand paging for out-of-core visualization." In *Proceedings of the 8th conference on Visualization'97*, pp. 235-ff. IEEE Computer Society Press, 1997.
[2]Manyika, James, Michael Chui, Brad Brown, Jacques Bughin, Richard Dobbs, Charles Roxburgh, and Angela H. Byers. "Big data: The next frontier for innovation, competition, and productivity." (2011).

However, it's all the same type of data. It's usually just the stock symbol and the price—mostly text. They collect transaction data and not pictures or sounds or news stories. So they don't have a big data problem. They certainly collect a lot of data, but the technology they have in place should be more than capable of handling the challenge.

Finally, think about the veracity of the data. Imagine you created a database that collected all the tweets and Facebook posts about your web site. You collect videos, pictures, and text. Several petabytes of data stream into your cluster every day. You run reports to see if the customers felt positive about your product. After you look through the data, you realize that there is not a question there to determine the customer's mood. All that effort was spent collecting useless data, because that data doesn't provide any information you need.

To provide an interesting example of a big data problem, think of the challenge surrounding self-driving cars. What type of data would you have to collect? You would need to collect massive amounts of video, sounds, traffic reports, and GPS location information—all flowing into the database in real time and at a high velocity. Then the car would have to figure out what data has the highest veracity. Is that person on the side road screaming because of a sports match or are they screaming because someone is standing on the road? A human driver has seconds to figure that out. A big data car would have to instantly process the video, audio, and traffic coordinates, and then decide whether to come to a stop or just ignore the sound. That's a real big data problem.

Tip Try to remember the difference between big data and data science. Big data will let you ask more interesting questions. That doesn't mean that all interesting questions need big data. Focus on the science. That way, whatever the data you have, you'll always be able to ask the best questions.

Summary

In this chapter, you learned that because data science revolves around interesting data, you often have to work with several types of databases. You learned some of the terminology used specifically for databases as well as some of the basic concepts and terms around the technology. You also saw how databases are organized. In Chapter 3, you'll learn how to recognize different data types.

Recognizing Different Data Types

When you're on a data science team, you'll often deal with many different types of data. These different types will be a key factor in determining how you'll want to store your data. Technologies like NoSQL provide you with a lot of flexibility to store different data types. Relational databases give you less flexibility, but they're sometimes easier to work with, and it's generally easier to generate reports in relational databases.

When you think about how you want to store your data, you need to understand the different data types. The same is true with any storage. Certain databases are optimized for certain types of data. Just like you wouldn't want to store a sandwich in a water jug, you wouldn't want to set up a relational database to hold the wrong type of data.

There are three types of data that your team should consider:

- **Structured:** Data that follows a specific format in a specific order.

- **Semi-structured:** Data with some structure, but also with added flexibility to change field names and create values.

© Doug Rose 2016
D. Rose, *Data Science*, DOI 10.1007/978-1-4842-2253-9_3

- **Unstructured:** Data that doesn't follow a schema and has no data model.

We explore each of these types of data in more detail in the following sections, and then cover what big garbage is and provide you some tips on sifting through it.

Keeping Things Simple with Structured Data

The first type of data is in many ways the simplest. It's commonly referred to as structured data. Structured data is data that follows a specific format, in a specific order. It's like the bricks and mortar of the database world—it's cheap, inflexible, and requires a lot of upfront design.

A good example of structured data is your typical office spreadsheet. When you fill your rows with data, you have to stick to a pretty rigid format and structure. For example, you may have a column called "Purchase Date." Each field has to follow a strict guideline. You can't put "Tuesday" in one row and then "March" in the next. You have to follow a specified format; for example, numerical month followed by a slash, day, and year (something like the MM/DD/YYYY format).

This format and structure is called the **data model**. Structured data relies on this data model. A data model is similar to a data schema, except the schema works to define the whole database structure. A data model defines the structure of the individual fields. It's how you define what goes into each data field. You decide whether the field will contain text, numbers, dates, or whatever.

Think about the spreadsheet example and what might happen if you ignore your data model. If you type Tuesday into the Purchase Date field in one row and March in another row, what happens when you want to create a report that displays all of the purchases in March? How would you do that? Would you use the number three? Would you use the word March? You certainly wouldn't use the word Tuesday.

If you did this type of data entry, your spreadsheet would be filled with data garbage. Every time you would try to sort the data or create a report, there'd be a bunch of rows with invalid data. Then you'd have to go back and clean it up or just delete them from the report. That's why many spreadsheet applications have formatting rules that force you to follow a particular model when you're entering data.

The same is true for databases. Many databases reject data that doesn't follow the model. Often the web site (or middleware) used to collect the data is set to a specific type and format for various fields.

Relational databases excel at collecting structured data, which means that there's a lot of structured data out there. A lot of the data that you access on web sites or through mobile apps comes from structured data. Your bank statements, flight information, bus schedules, and even your address book are all forms of structured data.

That doesn't mean that *most* data is structured. Actually, most data does not follow a specific format and structure. In fact, some of the more interesting data doesn't follow any structure at all. Data like videos, audio, and web pages have no defined structure.

As part of a data science team, you'll need to combine the type of data with the method of collection. If you use a relational database, you're limited to mostly structured data. With structured data, it's usually pretty straightforward to create reports. You can use the structured query language or SQL to pull data from your database and display it in a standard format. If you use a NoSQL cluster, you can work with all datatypes, but it will be more difficult to create reports. These are all decisions that your team needs to think about.

Sharing Semi-Structured Data

When you have structured data in your relational database, everything in the world seems defined and well organized. It's like when you have all of your spices in spice jars—you know where everything is and you know exactly where to find it. However, few applications ever stay that simple.

Semi-structured data is a little harder to define than structured data, so we'll use our running shoe web site as an example. Imagine you use a relational database for the running shoe web site. It has four tables: the shoes, the customers, their address, and shipping options. All of your structured data fits into a data model. The dates are standard and the ZIP codes are standard. Things are running smoothly. Everything seems right in the world.

Then you got an e-mail from your shipping carrier. The carrier says that you can dramatically lower your costs by adding information directly into their database. You just need to query their database, download one of the regional shipping codes, and then add it to the order and create a new record. It should be easy because their database is just like yours. It's all structured data and in a relational database.

The problem is that their schema is not the same as your schema. You called your ZIP code data "ZIP Code." They called their ZIP code data "postal code." You don't care if the shoes are shipped to a business or a residence. They do. You don't specify whether it's a house or an apartment. They have different rates for each. Now you need a way to exchange your structured data with their structured data, even though each of them is a different schema.

To solve that, you need to download the carrier's data and the related schema. When a customer orders a shoe, your database will send the ZIP code to the carrier's database. It will get back a bunch of data that includes their version of the address with their field names and their data model. Remember that they used the name "postal code" for ZIP codes. That will be included in the new data.

Their data has some characteristics of structured data. It is well organized and has a standard format. The text fields are text. The date fields are dates. But the data includes their schema. The carrier can use whatever names they want. That's why this type of data is called semi-structured data.

Semi-structured data is even more popular than structured data. It has structure, but that structure depends on the source. You'll work with semi-structured data all the time. Your e-mail is semi-structured data. It has a pretty consistent structure. You always have a sender and recipient, but the message may vary. The message contents could be just text or include images or attachments.

Data science teams typically work with more semi-structured data than structured data. There are volumes of e-mail, weblogs, and social network site content that can be analyzed.

There are a few terms that are fairly common when you're talking about working with and exchanging semi-structured data. One of them is the Extended Markup Language (XML) data type, which is an older semi-structured data type used to exchange information. There is also JavaScript Object Notation (JSON), which is an updated way to exchange semi-structured data. It's often the preferred data type for web services.

Including semi-structured data is a good way to ask more interesting questions. Back to the running shoe example. Let's say that you wanted to get customer feedback on your running shoe orders. You may download semi-structured data from some of the largest social media sites, and then combine that data with the structured data you have on your customer. If your customer is unhappy with their shoes, you can send them an apologetic coupon.

These are issues you can discover using structured and semi-structured data. Your team can start to investigate how happy your customer is with his or her purchase.

Collecting Unstructured Data

The most popular type of data is everything that isn't structured or semi-structured: unstructured data. Some analysts estimate that 80% of data is unstructured. When you think about it, this makes a lot of sense. Think about the data you create every day: every time you leave a voicemail, every picture you upload to Facebook, the OneNote memo or PowerPoint presentation you created at work, and even the data that's generated when you search on the Web. It is all unstructured.

So what does all this data have in common? That's the biggest challenge. The answer is not much. It's schema-less. Remember that the schema is the map that shows the data's fields, tables, and relationships. You don't have that map with unstructured data. In addition, format of unstructured data depends on the file. A Microsoft Word document might have a set format, but that format is only used by that application. It's not the format for all text. That's why you typically can't edit Microsoft Word documents in another program.

That also means there's no set data model. There's no consistent place to look for field names and data. If you had a dozen documents, how could you figure out their title and contents? What if some of them were PDFs, some were Microsoft Word documents, and some were PowerPoint Presentations? Each one has a proprietary format. There's no field to look up that has the label "document title."

This is a challenge that search companies like Google have been working on for years. How do you work with data with no set format and without a consistent data model? Every time you search these engines, you'll see the fruits of their labor. If you search for a term like "cat" you'll see text, videos, pictures, and even audio files.

Working with unstructured data is one of the most interesting areas in data science. New databases like NoSQL allow you to capture and store large files. It's much easier to store it all in one place. All that audio, video, pictures, or text files can go into a NoSQL cluster.

If you want to capture everything, there are new tools for that as well. You can use big data technology like Hadoop for processing data in batches or real time.

So let's go back to your running shoe web site. The business has grown a little, and now you're part of the new data science team. You work with marketing and management to come up with your first interesting question: who is the best running shoe customer? You gather some basic biographical information, which was pretty easy to find in your customer database. You have their e-mail address and the city and state where they live. You take that information and start crawling through your customer's social network posts. You start to gather all the unstructured data. Maybe your customer posted a video of finishing a marathon. You can send out a congratulatory tweet.

You might also decide to start crawling through your customer's friends' posts. Maybe your customer posts an image of them running with a group of people. You can use unstructured data to identify those people and send them special promotions.

This type of data project is typically called a **360° view** of your customer. You're trying to find out everything that you can about what motivates them. You can then use that information to find your best customers and send promotions. You may also find that you have a few customers who are referring a lot of their friends. You may want to offer them special incentives and discounts.

As time goes on, you can capture more and more of your customers' unstructured data, which will allow you to ask more sophisticated questions about your customers. For example: are they more likely to travel? Are they more competitive? How often do they go to restaurants? Each of these questions can help you connect with your customer and sell more products. As you collect this data, you may want to display it in a chart, as shown in Figure 3-1.

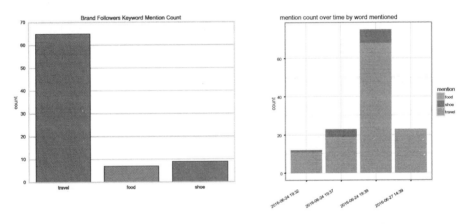

Figure 3-1. Brand followers keyword mention count

▓ **Note**   See how to create this chart at `http://ds.tips/Muj7f`.

Unstructured data is a resource that increases every day. Think about the things you did today that might be interesting to a company. Did you send a tweet about your long walk to work? Maybe you need better shoes. Did you complain about a rainy day? You should buy an umbrella. Unstructured data allows companies to offer that level of interaction.

Sifting Through Big Garbage

Unstructured data brings with it a new set of challenges. One of the first questions you run into is whether you want to delete data. Remember that a data science team uses the scientific method on their data. You want to be able to ask interesting questions. You need to decide if there's any limit to the questions you'd like to ask.

There are good arguments for both keeping and throwing away data. Some data analysts argue that you can never know every question you might ask, so why throw away data? It's also relatively cheap to keep massive amounts of data—often only a few cents per gigabyte. You may as well keep it all instead of deciding what to throw away. Sometimes, it's cheaper to buy new hard drives than it is to spend time in data retention meetings.

Other analysts argue that you should throw away your data. There can be a lot of garbage in those big data clusters. The more garbage you have, the more difficult it is to find interesting results, because there's too much noise (meaningless data) in your information. Deciding whether to keep or delete data is a struggle that many data science teams are still figuring out.

I once worked for a company that was dealing with this challenge. They owned a web site that connected potential car buyers with automobile dealerships. They created a tagging system that would record everything their customers looked at on their web site. Anytime a customer rolled over an image, the database would add a new record; all the links they clicked were collected. The system grew into thousands of tags. Each of these tags had millions of transactions. There were only a few people within the company that understood what each tag captured, which made it very difficult for them to create interesting reports.

They used the same tagging system with their unstructured data. They started collecting information on advertisements and Flash videos. They wanted to connect the tag to the image and the transaction, which allowed them to see the image that the customer clicked as well as the tag that indicated where it was located on the page. All of this information was stored in the growing Hadoop cluster.

Some people on the team argued that a lot of the data was obsolete. Only a few people knew the tagging system, and the advertisements constantly changed. In addition, the people who knew the tagging system started to rename the tags. So much of the data was obsolete.

Others on the team argued that this was a very small amount of data compared to what could be stored in the Hadoop cluster. Who cared if you had a couple of gigabytes of obsolete data? It wasn't worth the effort to clean up.

Chances are, you will deal with these types of challenges as well. As you do, keep these things in mind:

- There really isn't a right answer. Your data science team just needs to figure out what works best for them.

- If you decide to keep everything, you probably have to work a little harder when you're creating interesting reports. You will have to do a little bit more filtering and there will be a little bit more noise in your data.

- If you decide to throw data away, you'll have a cleaner cluster. However, there is a chance that you'll inadvertently throw something away that you may one day regret. It's like when you clean out your closet. You never know if that suede collar jacket will come back in style. But if you keep too many jackets, you may forget what you have.

The most important thing is to make sure that your team makes a decision. You don't want to have a data policy that changes every few months. Either decide at the beginning that you plan to keep everything or that you want to throw some data away. Work with the team to make sure that everyone agrees on the policy and what can be thrown away. If you don't have a set policy, you may corrupt all the data. If you don't know what you've thrown away and what you've kept, it's difficult to make sense of reports. Try to decide early on what works best for your team.

Summary

In this chapter, you learned that structured data is data that follows a specific format in a specific order. You also saw that semi-structured data is data with some structure, but there's added flexibility to change field names. Finally, there is unstructured data, which is everything else. It's data that doesn't follow a schema and has no data model. You also learned about big garbage and found out some tips for sifting through it. In Chapter 4, you will learn how to apply statistical analysis to your data.

Applying Statistical Analysis

Data science teams will spend most of their time collecting and storing data, and then using that data to ask questions. They create reports using statistics and math to see if they can get at answers. Statistics is a very interesting field. To participate in a data science team, you need some basic understanding of the language. There are several terms you need to be familiar with as you explore statistical analysis. They are:

- **Descriptive statistics:** The process of analyzing, describing, or summarizing data in a meaningful way to discover patterns in the data.

- **Probability:** The likelihood that something will happen.

- **Correlation:** A series of statistical relationships that measures the degree to which two things are related. It's usually measured as a number between 1 or 0.

- **Causation:** When one event is the result of the occurrence of another event.

© Doug Rose 2016
D. Rose, *Data Science*, DOI 10.1007/978-1-4842-2253-9_4

- **Predictive analytics:** Applying statistical analysis to historical data in an effort to predict the future.

We cover each of these in more detail in the following sections. As you're reading these sections, be sure to look for more statistical analysis terms.

Starting Out with Descriptive Statistics

Statistics are tools to tell a story, but they're not, in themselves, the end of the story. The best way to tell how much of the story you're getting is to push back when things don't seem right.

My son once told me a great joke about this that shows how teams can use statistics to tell stories. He asked, "Why don't you ever see elephants hiding in trees?" I shrugged and he said, "Because they're really good at it." Try to remember this joke when you look at your reports. People usually think of statistics as concrete mathematics. Who would question if two plus two equals four? In truth, statistics is much more like storytelling. Like any story, it can be filled with facts, fiction, and fantasy. You can hide some pretty big elephants if you don't know where to look.

One place you'll see this is with politics. One representative may say, "Over the last four years, the voters' average salary has gone up $5,000." People will clap. The challenger may say that they shouldn't be clapping and point out that "the typical middle-class family now earns $10,000 less than they did four years ago." Who's telling the truth? The answer is both of them. They're just using statistics to tell different stories. One story talks about prosperity and the other talks about failure. Both of them are true, and yet neither of them is telling the whole truth. You have to find the elephants in each of these stories.

In this case, the representatives are using descriptive statistics. They are trying to describe how all the voters are doing without talking to each family. They are creating a story of a typical family.

One representative uses something called the **mean**. This is one of the most useful and popular descriptive statistics. You see it used with grade point averages, sports scores, travel times, and investments. In this example, the representative adds up the income of every family, and then divides it by the total number of families. Sure enough, each family earned about $5,000 more.

But hold on. The mean is not the only way to describe a typical family. The competing representative has another way. She uses the median family income. The **median** describes what a family in the middle earns. To find this figure, you rank all of the families from lowest to highest, and then number them from top to bottom. You find the number in the middle by dividing the ranking in half. The family in the middle has the median income.

The competing representative finds the median income is $10,000 less. This indicates that the mean family income is $5,000 more, but the median family income is $10,000 less. These are two descriptions of the same families that have different stories, as shown in Figure 4-1.

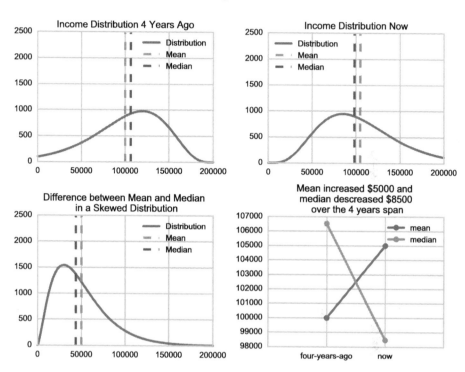

Figure 4-1. Different stories for the same families

Note   See how to create these charts at http://ds.tips/c6Aha.

When you see this, remember to look for that elephant. When there's a big variation between the median and the mean, it usually means that your data is skewed. In this case, imagine that a few of the families are extremely wealthy. In the last few years, their income has increased substantially. This accounts for millions of added dollars. These families would skew the data because there's a big chunk of money at the top. That would bump up the mean, but wouldn't really impact the median. In the mean, their income would be added up like everyone else's. In the median, they would probably just be at the top of the ranking. The family at the middle point wouldn't really change at all.

You'll see this challenge with the median and the mean in other ways as well. If there are two people standing in a room, their mean height might be just under six feet. If a basketball player walks into the room, their mean may increase by a foot. The median height would stay roughly the same.

On your data science team, don't be afraid to ask questions when you see stories told using statistics. Also, try to make sure that your reports use different ways to describe the data. Descriptive statistics can tell many different stories.

Understanding Probability

Probability is another area in statistics that allows you to tell interesting stories. The **probability** is the likelihood that something will happen. It's a measurement of the possible outcomes. If you flip a coin, probability gives the likelihood of the coin landing on one side or the other. The statistics side of probability focuses on probability distribution. If you throw a six-sided die, that means that there are six possible outcomes, and the possibility of any number coming up is one in six. That means each time you throw the die, you have about a 17% chance of hitting a particular number. Probability can also show a sequence of events. What if you want to show the likelihood of hitting the same number twice in a row? Well that's 17% of 17%, or roughly 3%. If you're playing a dice game, that's a pretty low probability. Your data science team will certainly want to work with probability. It's a key part of predictive analytics. It helps you figure out the likelihood that your customer will do one thing over another.

I once worked with a biotech company that was trying to use data science to predict the likelihood that someone would participate in a clinical trial. Getting people to participate in clinical trials is a tricky business. There are a certain number of clinics, and the company needs to keep them up and running—even if they're empty. If they don't fill up, the company loses revenue.

One interesting question they asked was, "What are some of the things that keep people from participating in clinical trials?" and it turns out that there are a number of things that might decrease the probability of someone participating. If people have to fast the night before, they may be 30% less likely to participate. In addition, if there are blood tests and needles, they may be 20% less likely to participate. Figure 4-2 is a flow chart that shows the three-way relationship between prior attendance, testing, and fear of blood.

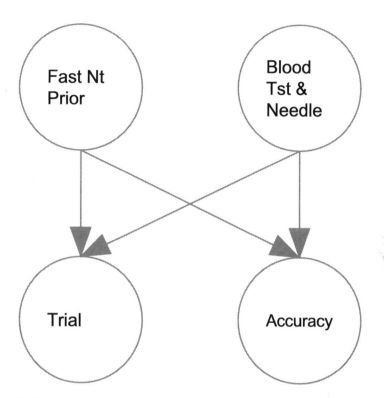

Figure 4-2. Three-way relationship between prior attendance, testing, and fear of blood

▨ **Note** See how to create this chart at http://ds.tips/V2tup.

The company had to balance out the probability of people participating against the accuracy of the results. For example, there was a drug trial, and the company could check the drug's effectiveness using either a saliva test or a blood test. The blood test is 10% more likely to be accurate. That was easy; they should just use the blood test. But hold on. If they run the trial with a blood test, they will have 20% fewer participants because of the people who decided not to do the study because they were afraid of needles. If they want 1,000 participants, that means about 200 fewer people.

That brought up another interesting question. If the test has 200 fewer people, does that mean that they'll have a less accurate drug trial? The data science team created another probability distribution. What if the drug has a 1/500 chance of causing some type of reaction? You'd have a much better study with 1,000 people than you would with 800.

The data science team had to take that into account. Was it better to have more people participate in the study without needles even though the saliva test was less accurate? This led to even more questions. Should the team have participants take the saliva test several times to increase the probability of an accurate result? In the end, that's what the data science team helped the company decide. Maybe it was best to have the greatest number of people participate in the trial to increase the probability of catching a drug reaction. Then, they could perform the less accurate test more often to increase the probability of having an accurate result. This would allow the company to have maximum participation and, at the same time, increase the accuracy of their study—all brought to you through the power of probability.

There are a couple things to keep in mind when you're working with probability on your data science team. The first is that probability will lead you to some unexpected places. Who would've thought that a medical practice might get better results by administering a less accurate test? The second is that probability can also be a great vehicle for asking interesting questions. Don't be discouraged if your questions just lead to more questions.

Remember that data science involves applying the scientific method to your data. Sometimes the path will lead you to unexpected places. The important thing is to not jump off when the path takes a sharp turn, which can easily happen when you're working with probability. Those sharp turns are often the path to your greatest insights.

Finding a Correlation

Correlation is another very interesting area in data science. Many companies use it to guess which products you'll buy. It's also used to connect you to friends and acquaintances. If you use a service like Netflix, you've probably been amazed at how well the web site can guess what movies you'll like. Amazon has been using correlation for years to recommend books.

Correlation is a series of statistical relationships that measures the degree to which the two things are related. It's usually measured as a 1 or 0. There's a correlation of 1 if two things are strongly related. There's a correlation of 0 if two things have no relationship. The 1 can be expressed as a positive or negative number. A -1 is typically an inverse or anti-correlation.

A positive correlation might be something like height and weight. A person is likely to weigh more if he or she is taller. As the height increases, so does the weight. There are even more straightforward examples, such as the higher the temperature is outside, the more people will buy ice cream. As the temperature goes up, ice cream sales go up. A negative correlation might be something like cars and gasoline. The heavier the car is, the less likely it is to get good gas mileage. As the weight of the car goes up, the gas mileage goes down. They have an inverse relationship.

If you're a runner, you might notice that you run slower as you go uphill. That too is a negative correlation. The steeper the incline, the slower you run. As the incline goes up, your speed goes down.

Both positive and negative correlations are great ways to see the relationship between two things. A negative correlation isn't bad. It's just another type of relationship.

A data science team will look for correlations in their data. They will try to fine-tune any relationship between people and things. Fortunately, software tools can handle a lot of mathematics behind calculating a correlation. One typically used formula is the **correlation coefficient**, which is the 1, 0, and -1 that indicate whether there's a statistical relationship between people and things.

When figuring the correlation coefficient, you typically won't get a nice, neat round number. Instead, you'll probably discover a .5 or a -.75 correlation. This indicates a stronger or weaker correlation—the closer you are to 1 or -1, the stronger the relationship.

One interesting data science challenge was LinkedIn's "People You May Know" feature. LinkedIn wanted a way to figure out when professionals knew each other. There are data science teams who work with LinkedIn data, look for correlations between connections, and then try to figure out why they're connected. The connections can be because of the schools they've attended, jobs they've shared, or groups and interests they've shared.

This data science team looks for positive and negative correlations. They might find information on the web site that shows that you work at a certain company and that you're interested in a certain job. Then they find someone else who is also interested in the same job and worked at the same company. That's enough to establish a correlation between you and this other person; therefore, the web site might recommend that you make a connection with this other person.

The data science team can also make a correlation between your connections and other people's connections. If you're connected with one person and they're connected with someone who has similar skills as you, you might make a good connection. If you think about it, this makes a lot of sense. You're much more likely to know people who work in the same office building. You're also more likely to be connected with people who have similar interests and skills. As the number of similar skills increases, the likelihood of you knowing that person also increases.

Correlation also has the power to help your team question its assumptions. You might assume that the people who spend the most money on your web site will also be your happiest customers. That might not be the case. In fact, there might be a negative correlation between the two. Maybe the people who spend the most actually have the most unrealistic expectations. They're easier to disappoint and more likely to leave negative feedback, as shown in Figure 4-3.

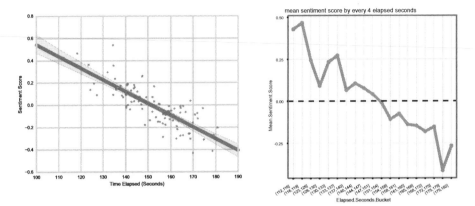

Figure 4-3. Sentiment score over time

 Note   The more time a visitor spent on the web site, the lower the sentiment score. See how to create this chart at http://ds.tips/pawR7.

As a data science team, you will use correlation to test your assumptions. You may look for strategies to get your happiest people to spend more. You may also look for ways to manage your high spenders' expectations. If you look for these correlations, you'll see a lot of things that you might otherwise miss.

Seeing That Correlation Does Not Imply Causation

Correlation is a great tool—it helps you see relationships that you might not otherwise see. But there's a flip side. You'll have to see whether the correlation caused something to happen. Was that the thing that started the change? As a general rule, correlation doesn't imply causation. A third thing that hasn't been considered might actually affect a relationship between two things. It's a big challenge for data science teams to figure out causality. You don't want to create relationships that don't exist.

Think of it this way. I grew up in a colder area of the country. When my parents got older, they moved to southern Florida. They now live happily in a sunny retirement community. Every few months, my family goes to Florida to visit. Statistically, their community is one of the most dangerous places on earth. Every time we visit, there are people being hospitalized or worse. There's a strong correlation between their community and death or severe injury. You'd think that because of this I'd never visit my parents. It sounds likes the opening scene of every first-person-shooter video game.

Yet we see past this correlation. We visit them often and it feels perfectly safe. It's because the correlation doesn't imply causation. The true cause is that the median age is much higher. Older people in a retirement community have a higher probability of injury or death. If you looked at this correlation, you'd think they lived in a war zone. You'd never imagine them peacefully playing mahjong by the pool.

Think about how your data science team might also apply these concepts. Let's go back to our running shoe web site. Imagine that the team identified that there was a big increase in sales in January. There's a strong correlation between January and the number of people buying new shoes.

The team gets together to understand the cause. They ask some interesting questions. Do people have more money in January? Are more people running during the coldest months? Are these first-time runners? Are they new customers? What kind of shoes are they buying?

The team discusses the questions and decides to create reports. The reports suggest that most of these customers are new customers buying expensive shoes. Because of these reports, the team feels comfortable that the cause of the new sales is that new customers have more money in January. Maybe they received gift cards.

The following year, the team decides to take advantage of this causation. In December, they offer discounted gift cards. They also send promotions to last year's new customers. A few months later, the team looks over the data. They find that their promotions and discounts had no impact. Roughly the same number of people bought the same number of shoes. Even worse, the last year's new customers seem to have no interest in new running shoes. Therefore, having more money *wasn't* the cause of the correlation. The data science team went back to their original questions, and ran a few more reports. They found that all of the new sales for both years were for new customers and first-time runners. Why would there be a burst of new customers buying expensive running shoes during the coldest months?

The team thought about it and considered the reason might be behavioral. They posed a new question. Are all the new customers people who are trying to get in shape because of a New Year's resolution? They went back and created reports, as shown in Figure 4-4. The reports said that they were all new customers that bought one pair and then stopped visiting the site in the middle of the year. This suggested that they bought expensive shoes and then gave up. The team guessed that the expensive shoes might've been a motivation to keep running. The next year they decided to create a new promotion. It was geared around New Year's resolutions. They sent out a mailer that said, "Do you want to keep your New Year's resolution?" It offered free running guides and fitness trackers as a way to keep people interested.

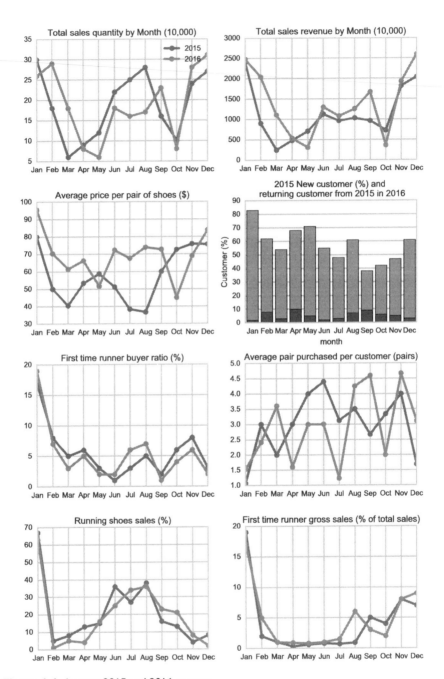

Figure 4-4. January 2015 and 2016 reports

▨ **Note**   January of 2015 and 2016 had the highest and second highest sales quantity and gross revenue. In both years, January had the average highest price per pair of shoes. The company offered more gift card discounts in 2016, but that did not boost the sales. The numbers of returning customers in 2015 and new customers in 2016 were the lowest amongst all the years. Around 15 to 20 percent of buyers in January 2015 and 2016 were first-time runners. On average, they bought 1 to 1.5 pairs of shoes. When you drill down to see the first-time runner segment customer behavior, the gross sales peak in January. Some of them come back in the fourth quarter, but about half of them never came back. See how to create this chart at http://ds.tips/fe2Ax.

Correlation and causation are key challenges for most data science teams. It's a real danger to create false relationships. In statistics, this is called a **spurious causation**. As you can see, finding the real cause will give you much greater value. The best way to avoid spurious causation is by following the scientific method. Remember to ask good questions and be clearheaded about the results.

Combing Techniques for Predictive Analytics

So far, everything you've seen has been about the past. You've seen how to collect different data types and then perform statistical analysis. These statistics were a starting point for greater insights. Your data science team will start to create correlations and see the probability of certain events. Now, let's take these ideas and turn them around to predict the future—typically called **predictive analytics**. This term is closely associated with data science—so much so they're sometimes even used interchangeably—however, they are not the same. Predictive analytics is a type of data science. Data science is applying the scientific method to your data and predictive analytics takes that result and makes it actionable.

Think of it this way. Meteorology is a type of science. These scientists study physics, wind speed, and the atmosphere. If you're outside with a meteorologist, he or she will explain to you why the clouds look a certain way and how the pressure determines their movement. That's the science side of meteorology. It's about understanding the weather and seeing historical trends. The more common view of meteorology doesn't revolve around the science; rather, it has to do with weather forecasting.

Weather forecasting is when a team of meteorologists uses predictive analytics. They can use historical data to assign probabilities and use probability and correlation to predict weather patterns. There might be a correlation between low pressure systems and severe storms. As the pressure decreases, the severity of the storms increases. There's a positive correlation between pressure and storms.

All this analysis comes together so that the meteorologists can answer a simple question: what's the weather going to be like tomorrow? What was once about understanding the past now becomes a prediction for the future.

Currently, there's a growing interest in predictive analytics because new tools and technology allow for more interesting insights. Think about weather forecasting. Right now, the weather service is restricted to historical data from a few thousand stations. Imagine that the weather service gave out millions of low-cost beacons. People would install them in their homes and connect them to their wireless networks. These devices would record air pressure and temperature information, as well as video and audio, and then upload data to a national Hadoop cluster. This would give the scientists unprecedented levels of information.

That's why predictive analytics is so closely associated with data science. The higher volumes of data allow the team to ask interesting questions. Then the team can perform complex analysis. Here, the team would be able to look at weather patterns house-by-house and block-by-block, and then create complex predictive models based on millions of homes.

The same holds true with your team. Think about the running shoe web site. Imagine that your team collects millions of tweets about running. The team identifies a few influential runners on Twitter. You could then send them free shoes or promotions hoping they would say nice things about your company. You could also use this data to identify key running events.

These new tools allow data science teams to achieve a much larger view of data as well as look at waves of information in real time. Meteorologists can analyze petabytes of information, and the running shoe web site team can analyze millions of tweets.

Organizations usually get really excited about the idea of predictive analytics—so much so that they don't always put enough time into developing their data science teams. They want to go straight to predictions without understanding the data they already have. When you're working on a data science team, be sure to communicate that the quality of the predictions depends on how well the team has analyzed the data.

Your team has to understand the past to be able to predict the future. Don't shortchange your analysis. Ask good questions of your data and use your statistical tools to create interesting reports. Once you do that, your predictions of the future are far more likely to be accurate.

Summary

In this chapter, you learned about descriptive statistics, probability, correlation, causation, and predictive analytics. You also learned how to use statistics in your data science stories. In Chapter 5, you'll learn how to avoid pitfalls when getting started with data science.

Avoiding Pitfalls in Defining Data Science

Throughout this book, I include this Avoiding Pitfalls chapter at the end of each section to help you apply these ideas to your organization. There are often hiccups between new ideas and how things will play out in most organizations. You might not run into these challenges, but they are common to many organizations.

Focusing on Capability

One of the key challenges in data science is what I call the "cluster of dreams." It's based on the movie *Field of Dreams,* starring Kevin Costner as a man who spends his life savings building a baseball diamond in a cornfield. The ghosts of old players visit him and tell him to finish the baseball diamond. They say, "If you build it, they will come."

Many organizations get caught in the same trap. They focus their energy on building hardware and collecting massive amounts of data. They make large investments in software to run on large data clusters. Their dream is that if they have enough hardware and software, they'll gain valuable insights. (If they build it, they will come.)

© Doug Rose 2016
D. Rose, *Data Science,* DOI 10.1007/978-1-4842-2253-9_5

This makes a lot of sense when you think about it. Many organizations have a lot of experience delivering successful hardware projects. It's something they know how to do. They're good at it and they've done it for decades. Hardware is real—you can see what you're getting. Exploration is tougher to quantify. It doesn't have a return on investment that fits neatly into a project pipeline. You only know if it was worth it after you've already done it. Data science is new, and in many organizations, it's not easy to spend money on exploring and asking questions.

The Library of Congress famously started a project to collect 170 billion tweets. They wanted to show that they could work with data science. They bought the hardware to store the tweets, but they didn't have any plan for what to do with the data. They also couldn't give anyone access to the data. They figured if they built it, then they would come. Unfortunately, the data sits idle on hundreds of servers. It was a monument to data collection.

This might seem like an extreme case, but it's very common. Organizations focus on building out capability. They set a goal for setting up a certain number of nodes in their Hadoop cluster and using a suite of visualization tools. The budget goes into the hardware and software, and there is little left over for the data science team.

I once worked for an organization that was trying to use a big data cluster to replace their data warehouse. They were used to spending millions on hardware and software, and then they would hire warehouse experts to maintain their investment. When they moved to Hadoop, they had that same mindset. They started a multimillion-dollar project to create three separate clusters. The entire budget went into servers and software.

At the end of two years, they had three clusters, but only a few people knew how to access the data. To make matters worse, these people were spread across several different functional areas. They had millions of dollars invested in hardware and software, but no data science team to analyze the data and gain insights. A couple of years into the project, the cluster had just a few terabytes of data—roughly the same amount you could fit on a hard drive for a few hundred dollars. Only a few people accessed this data to create a few simple reports for one or two departments.

There are a few things to remember to keep from falling into this trap. The first thing is that data science teams are exploratory. They're looking at the data to find insights. Data isn't the product; it's the insights that come from the data. There's no prize for having the biggest cluster.

Even though a data science team might spend most of their time collecting data, that doesn't mean that all your value comes from the collection. It's the same way that having a chef's knife doesn't make you a chef. A big data collection doesn't make you a data science team. It's the questions you ask and the creation of organizational knowledge.

Most data science teams will use several different software tools. (Sometimes they'll want to use R instead of Python; it might be easier to hold a small subset of data in a relational database like MySQL; and they might use different visualization tools). Give your team the option to be flexible and explore. Often, a data science team can get a lot more done with several free tools than they can with one big investment.

The science team should build out tools as they need them. A good data science team will be messy; they will use many different tools and techniques to wrangle and scrub their data.

Instead of hardware and software, invest in training and expertise. The most valuable part of your data science team is the people asking interesting questions and delivering new knowledge.

Summary

In this chapter, you learned that it can be a mistake for organizations to focus on the capability. Companies should focus more on the training and expertise of their people rather than the hardware and software required to gather the data. In Part II, you will learn how to build your data science team, starting with rounding out your talent in Chapter 6.

Building Your Data Science Team

It's time to build your data science team. Before you can build it, you need to understand what types of team members you need, how to form the team, what kind of work your team members will do, and how to work together. Then, as always, you should understand how to avoid any pitfalls you may encounter.

Rounding Out Your Talent

We defined data science in Chapter 2 and covered what it means to be a "data scientist." In this chapter, you'll see how to break that role into several team roles. Then you'll see how this team can work together to build a greater data science mindset.

Putting Data Scientists in Perspective

As you learned in Chapter 2, there's some confusion surrounding the role of a data scientist. In 2001, William S. Cleveland published "Data Science: An Action Plan for Expanding the Technical Areas of the Field of Statistics."[1] This paper was the first to merge the fields of statistics and computer science to create a new area of innovation called "data science." At the same time, Leo Breiman published "Statistical Modeling: The Two Cultures,"[2] which described how statisticians should change their mindset and embrace a more diverse set of tools. These two papers created a foundation for data science, but it built on the field of statistics.

---

[1]Cleveland, William S. "Data science: an action plan for expanding the technical areas of the field of statistics." *International statistical review* 69, no. 1 (2001): 21-26.
[2]Breiman, Leo. "Statistical modeling: The two cultures (with comments and a rejoinder by the author)." *Statistical Science* 16, no. 3 (2001): 199-231.

© Doug Rose 2016
D. Rose, *Data Science*, DOI 10.1007/978-1-4842-2253-9_6

In 2008, some top data gurus from Facebook and LinkedIn got together to discuss their day-to-day challenges. They realized they were doing similar things. They saw their role as a crossover of *many* different disciplines. They decided to call this role a "data scientist."

A data scientist at this time was just a list of qualities. For example:

- Understand data

- Know statistics and math

- Apply machine learning

- Know programming

- Be curious

- Be a great communicator and hacker

They were renaissance enthusiasts who crossed over into many different fields.

The problem is that this list of skills is not easily found in one person. Each of us is predisposed to certain areas based on our individual talents. We usually gravitate toward our talents, and then work to refine our craft. A statistician will often work to become a better statistician. A business analyst will work to refine his or her communication skills. There is also a lot of organizational pressure to specialize. Most large organizations are divided into functional areas. There's some need for common understanding, but not always common expertise.

People are also notoriously bad at self-assessing their own abilities. The famous Dunning Kruger study[3] found that people who rated themselves as highly skilled often dramatically overestimated their expertise. A gifted statistician may rate themselves as an excellent communicator, but you don't need to be a good communicator to be a great statistician. A great statistician could easily have a long career even if he or she fumbles through presentations.

That's why most organizations divide the work up into teams. Individuals on the team will have their own areas of expertise. A cross-functional team doesn't assume that everyone is an expert. Instead, it encourages individuals to learn from each other's strengths and cover each other's weaknesses. A team of data scientists might not be able to identify those weaknesses. The team will blindly fumble if there's no one to identify blind spots.

---

[3]Kruger, Justin, and David Dunning. "Unskilled and unaware of it: how difficulties in recognizing one's own incompetence lead to inflated self-assessments." Journal of personality and social psychology 77.6 (1999): 1121.

I once worked for an organization that had a team of data scientists building out a cluster. There was some concern from the business because the higher-ups had no idea what the team was building—they were frustrated because they were paying for something they didn't understand. I went to a few of the meetings. The team of data scientists demonstrated a simple mapReduce job. The business managers stared blankly at the screen and occasionally glanced at their smart phones. To an outsider, it seemed obvious from the yawns and eye rubbing that the team was not doing a great job communicating.

After the meeting, I wrote a matrix on the whiteboard. I listed the following six skill sets:

- Data
- Development
- Machine learning
- Statistics
- Math
- Communication

I asked the data scientists to rate how they felt they were doing on each of these areas from 1 to 10 (1 being poor and 10 being best) so we could look for areas to improve. I took that same list of skill sets and showed it to one of the business analysts. I asked them to rate the team.

The results are shown in the Table 6-1.

Table 6-1. Data scientists' and business analysts' ratings

Skill Set	Data Scientists' Ratings	Business Analysts' Ratings
Data	8	10
Development	7	9
Machine learning	6	8
Statistics	8	9
Math	8	10
Communication	9	6

It was a classic Dunning Kruger result. In the places where the data scientists rated themselves as highly skilled, they dramatically overestimated their expertise. The data scientists all came from quantitative fields. They were statisticians, mathematicians, and data analysts. They couldn't identify their own blind spots. It took someone from an entirely different field to shine a light on their challenges.

If you're part of a large organization trying to get value from data science, it would be a mistake to rely on a few superhero data scientists. Individuals who come from a similar background have a tendency to share the same blind spots. Academic research shows that you often get better insights from a cross-functional team with varied backgrounds.[4]

There is some wisdom in our eclectic organizational structures. People with marketing, business, and management backgrounds deserve their place at the data science table. It's unrealistic to assume that key people with a quantitative background will have all the same questions and insights. Keep your team varied and you're more likely to have great results.

Seeing Value in Different Skills

One of the dangers to your data science team is putting too much emphasis on data scientists. Remember that data scientists are multidisciplinary. They should know about statistics, math, development, and machine learning—all while understanding the customer and coming up with interesting questions. Most data scientists come from engineering, math, and statistics backgrounds. This means that they're likely to share a similar approach to questioning and look at the data from a shared perspective.

It's unlikely that someone who's spent a career in math and statistics will have as much insight into customers as someone who's spent his or her career in marketing. Being an expert in one field doesn't assume expertise in another.

Many people who claim to be multidisciplinary usually have a few very strong skills with some knowledge of other areas. If you're very confident in many areas, you're likely to have large skill gaps. It also means that a team that only has data scientists can have similar blind spots and be prone to groupthink.

One way to keep this from happening is to allow people with other backgrounds to participate in your data science team. Remember that good data science relies on interesting questions. There's no reason why these interesting questions should only come from people who analyze the data.

Think about your running shoe web site. A data analyst shouldn't have a problem finding web sites that referred customers to the store. Let's say that most of the customers came from Twitter, Google, and Facebook. There were also quite a few customers who came from other running shoe web sites. A good data analyst can easily create a report of the top 50 web sites customers visited just before buying from you. Trying to find out where people are coming from is a good analytics question. It's about gathering the data, counting it, and displaying it in a nice report, as shown in Figure 6-1.

[4]Bosquet, Clément, and Pierre-Philippe Combes. "Do large departments make academics more productive? Agglomeration and peer effects in research." Spatial Economics Research Centre Discussion Paper, no. 133 (2013).

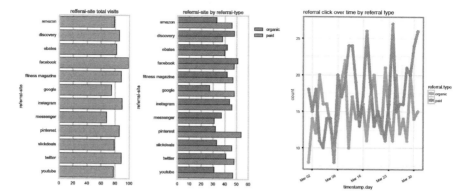

Figure 6-1. Referral-site total visits and referral type

Note Facebook, Twitter, and Instagram seem to bring great traffic in both paid and organic traffic. Pinterest drives a comparable amount of traffic to other sites, but about half of the traffic comes from paid advertisements. See how to create these charts at http://ds.tips/fRa4a.

A data science team goes deeper. The team might ask, why are there more people coming from Twitter than Google? Are people tweeting pictures of shoes? How many more people would visit the site if we bought advertising on Twitter? Is one site better than the other for releasing new products? Are people more likely to visit the site if they see a picture of a shoe? These questions are separate from the data. There's no reason a business analyst, marketing specialist, or project manager can't ask these questions.

A study of economics departments showed that when people from different disciplines collaborated, they were more likely to produce higher-quality publications. Diversity of opinion was a benefit to the quantity and quality of their work. In addition, people from different backgrounds are more likely to disagree. Disagreement causes everyone to work harder. In the end, this makes everyone's arguments stronger. If everyone on your team easily agrees on the best questions, you're probably not asking very interesting questions.

When you create your data science team, try to include many people from different parts of your organization. You want everyone in your organization to think about how they'll be more data-driven. If you only hire data scientists for your team, you're likely to make data science seem like a dark art—something only a few highly skilled people should attempt. This will make your data science less creative and disconnected from the rest of the organization.

In your data science team, it's important to separate analysis from insight. A data analyst captures, counts, and presents the data. Insights are much tougher to get. You need to follow the scientific method of posing interesting questions and looking for results. Don't let your team only produce data analysis. You want them to work harder. It's likely that someone from the business side will push the team to ask more interesting questions. It's also likely that someone from a marketing team will have interesting questions about your customer.

Some organizations have started moving in this direction. Companies like LinkedIn have created data walls that show different reports and charts from the data analysts. These walls of information allow people from all over the organization to see if there is anything interesting in the data. A marketing assistant might see an interesting story or an intern in human resources might think of an interesting question. This is a good way to get feedback from other parts of the organization.

Some organizations are going further and making sure that each data science team has a representative from both the marketing and the project management offices. This ensures that your data science team has someone who specializes in thinking about the customer as well as someone who understands how to deliver value to the rest of the organization.

Creating a Data Science Mindset

One term you'll hear frequently in relation to data science teams is "**data-driven**." It's a little bit of a tricky term. We all like to use data to drive our decisions. If you decide not to eat sushi from a gas station, it's based on real data. You're using past experience and maybe some observations to make a good decision. More often than not, your intuition is right—or at least half right. Try not to think of data-driven decision-making as a drop-in replacement for your own intuition. A data-driven culture uses data to enhance the team's intuition, not to replace it.

Your data science team will be the starting point for creating a larger data-science mindset that has a deeper relationship with data. Try to think of data-driven organizations as companies with many data science teams reinforcing a data science mindset. These teams create a culture of questions and insight. They should help the organization not only collect data, but also make it actionable.

A data science team will have three major areas of responsibility. These three areas create the foundation for your data science team, which will help the rest of your organization embrace this new mindset. They are:

- **Collecting, accessing, and reporting on data (groundwork):** This involves processing the raw data into something that everyone else can understand.

- **Asking good questions:** This drives interesting data experiments, and may come from the team members who don't necessarily have a technology background. They can be from business, marketing, or management. They ask interesting business questions and push everyone to question their assumptions.

- **Making the data actionable:** This will be the responsibility of team members who are primarily concerned with what the team has learned and how this data can be applied to the organization.

I once worked for a retail organization that sold home hardware and construction supplies. The company maintained several call centers because many customers preferred to call in their orders instead of using a mobile application.

The company was just starting out with data science and wanted the data science team to understand why these customers preferred to call in, because it costs a lot to maintain call centers. In addition, orders taken over the phone were much more likely to have errors. There were three people on the data science team: someone who understood the data, a business analyst, and a project manager. The three of them got together and tried to understand why these customers preferred to call in.

The business analyst was the first person to start asking questions. Do these customers have an account to order through their mobile phone? Are they professionals or residential customers? How much are they spending?

The team then created the data reports, shown in Figure 6-2. The data showed that most of the people were professionals who regularly placed several orders through their mobile devices. The orders they placed via the call center were much smaller than the orders placed through the mobile application. Around 80% of the transactions were less than $20. The business analyst had the follow-up question, "Why are some of our most loyal professional customers calling in for orders less than $20?"

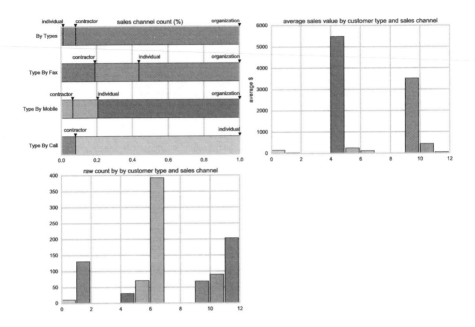

Figure 6-2. Data reports for sales channels

▨ **Note** Most of the orders are placed by organizations; however, most of the orders placed through calls are by individuals. The average total value of an order placed through calls by individuals is the lowest across all categories. See how to create these charts at http://ds.tips/3uprU.

After looking at the data and talking to a few customer service representatives, they figured out that these customers were calling because they needed a small part to fix a big problem. The customer service representatives were looking up that part while these professionals were on job sites. Most of the time on the phone was spent describing, identifying, and expediting a crucial part that they needed.

The team tried an experiment. They contacted a few of their high-volume, professional customers and asked them to send a picture when they needed an emergency part. They called it their "Pic-it-Ship-it" program. They hoped this would increase customer satisfaction and decrease the time spent on the phone trying to describe the part.

The data science team was small, but they still covered all three areas of responsibility. They collected the data and created interesting reports. The business analyst asked interesting questions and got some insight into the customers. Finally, the project manager organized an experiment and started a small trial program. They collected new data, asked interesting questions, and made the insights actionable.

Before the data science team ran these experiments, the organization always assumed that these people were small-dollar, residential customers who were more comfortable on the phone than with a mobile application. Their intuition was only partially right. The majority of the people calling were actually some of their most valuable customers. A data science mindset with a diverse team led to better insights.

Summary

In this chapter, you explored the roles in a data science team. You found out what skills to bring to the table. You also saw how you can create a data science mindset. In Chapter 7, you'll find out how to form your team.

Forming the Team

Having a data science mindset is both an organizational and cultural challenge. It's not as simple as hiring a few data analysts. You want your entire organization to think about your data in creative and interesting ways. Data analysts will help you analyze your data, but they may not be the best source of new insights. As mentioned in Chapter 6, you should think about data science as a team endeavor—small groups of people with different backgrounds experimenting with your data to create knowledge. That's the scientific method in data science. It's an empirical process of exploration. You'll ask good questions, gather evidence, and try to draw conclusions.

Instead of finding a few people who can do everything, work with your existing people who can do most things. One model that I think works well is breaking down your data science team into three roles:

- Research lead

- Data analyst

- Project manager

I've seen these three roles at different organizations. They may have different names, but they look to accomplish the same thing: ask good questions, gather evidence, and try to gain insights. We cover each of these roles in more detail in the following sections, and then we talk about how they work together in a group.

© Doug Rose 2016
D. Rose, *Data Science*, DOI 10.1007/978-1-4842-2253-9_7

Asking Good Questions with a Research Lead

Albert Einstein was credited with saying that if he had an hour to solve a problem, he'd spend the first 55 minutes trying to find the right question. Questions are the key to discovery. That's what makes them essential for a data science team. Questions are the most important drivers for your new insights. The key part of "science" in data science is finding the right question.

You already have the data. Your organization is probably collecting more data than you need. In many ways, you've already gathered the answers. Now you have to come up with the right question. For most organizations, that's not an easy task. We still work for companies that crave answers because answers end the discussion. You get out of meetings and you start to put something into practice. That's why organizations tend to favor people who are experts. They supply the answers. Answers are final and close-ended; questions are open-ended. In addition, a good question can lead to more questions.

That's why data science teams should rely on someone who is solely focused on asking questions. This person should understand the business, but he or she also needs to step outside of the business. One of the best names I've heard for this role is **research lead**. It captures the person's primary responsibility: to lead the questioning and drive the research.

The research lead should be someone from the business side who pushes the team to ask interesting questions. She should start out by working with the team to come up with questions or identify key problems. She can put them on a question wall or organize them into sticky notes.

A research lead has three areas of responsibility:

- Identify assumptions
- Drive questions
- Know the business

These three areas are closely related. As a research lead, sometimes you have to step outside of your experience and approach the business like you're seeing it for the first time. This takes some effort, and you have to dedicate yourself to taking a fresh perspective.

A good research lead will know intuitively when there's an interesting question. This is what happened with Dr. Jason Steffen.[1] He was an astrophysicist and a frequent traveler. He found himself in a long line waiting to get on an airplane. Most people just accept this as part of traveling. Dr. Steffen understood the business of flying enough to realize that having passengers wait in a long

---

[1]Steffen, Jason H. "Optimal boarding method for airline passengers." *Journal of Air Transport Management* 14, no. 3 (2008): 146-150.

line was inefficient and costly. He asked the simple question, "Is there a better way to board an airplane?" His background in science helped him come up with a solution. He imagined people boarding in parallel, skipping every other row. This was a lot more efficient than the current method of having people line up one at a time. (Unfortunately, this practice has not been put into place for a variety of reasons.)

Dr. Steffen worked through all three actions of a good research lead. He had some knowledge of the business. He knew about flying. He was able to question his assumptions. Millions of people lined up before him without thinking twice about the boarding process. Finally, he came up with a simple question: is there a better way?

You don't have to be a scientist to come up with interesting questions. A research lead should have some idea of the business, but this person doesn't need to be an expert. For example, Edward Land invented the Polaroid instant camera because he was inspired by a question from his three-year-old daughter.[2] When they were on vacation in New Mexico, he took a picture with a conventional camera. His daughter simply asked, "Why do we have to wait for the picture?" His daughter questioned a simple assumption.

Both of these questions started interesting avenues of exploration. Dr. Steffen came up with a way for people to board a plane up to 30% more efficiently, and Edward Land created an instant camera, one of the most innovative and beautiful technology products of the 20th century.

One of the most helpful components of the research lead role is that it separates the questioning from the data. There's nothing inherently technical about asking good questions. Remember, Edward Land's daughter was only three years old.

There's also a lot of benefit in separating the person who drives the questions from the person who looks for possible answers. There's an inherent conflict of interest. If you only have a small data set, you might limit yourself to simple questions—ones you can already answer with your data. A good research lead might make you rethink the type of data you collect. In the end, that's much more valuable.

Presenting Data with a Data Analyst

Your team needs good data analysts. The data analyst is responsible for making sense of the data, obtaining and scrubbing the data, and then displaying the data in simple reports. They should work with the research leads to see

---

[2]Mintzberg, Henry. "The fall and rise of strategic planning." *Harvard Business Review* 72, no. 1 (1994): 107-114.

if anything jumps out of the reports. They should also recommend statistical methods or create data visualizations. The research lead and data analyst will work hand-in-hand to build insights. The research lead focuses on the best questions to ask, while the data analyst tries to provide the best reports.

There's a lot of confusion around the different data job titles. There are statisticians, statistical analysts, data analysts, data scientists, data engineers, and even mathematicians. You can even arbitrarily add "chief" or "senior" to any of these. A chief statistician might be much more senior than a senior data scientist. The reason for all the confusion is that people in these jobs all do something very similar. In one way or another, they all practice the science of learning from data; they just arrived at the role in very different ways.

Statistics has been around for hundreds of years. The discipline evolved because governments needed to understand their own demographic and economic data. It has a long and rich history. The American Statistical Association (ASA) is one of the oldest professional societies in the country.

Data analysts, on the other hand, come from computer science. They learn to extract meaning from relational and NoSQL databases. They focus on presenting and discovering interesting bits of data that support decision-making.

Data scientists are seen as multidisciplinary. They're data analysts, but they also create software, work with mathematics, know the business, and ask interesting questions. As the former chief scientist at Bitly, Hilary Mason created a popular definition. She sees data scientists as data analysts who also know math, software development, engineering, and hacking. They can take the next step. Instead of just producing reports, they can start searching for insights.

A data science team splits up the responsibilities that would typically fall on one data scientist. Often, it's simply too much to ask one person to understand the data and the business and ask interesting questions. A good data science team needs a data analyst who also knows a little bit about software development, and most data analysts already find it necessary to know about software development. Many great visualization tools require some software coding. Python and R are two of the most popular languages for exploring and displaying your data. (You learned some about these languages in Chapter 1.)

A data analyst has three main areas of responsibility on a data science team: preparing the data, selecting the tools, and then presenting the results.

Preparing the Data and Selecting the Tools

Preparing the data and selecting the tools go hand in hand. You have to select the tools to prepare the data. Therefore, as a data analyst, you spend most of your time preparing the data. You have to figure out the best way to obtain data—whether through web APIs, scraping it from pages, or collecting it from different parts of the organization—and then scrub the data. Scrubbing data makes it more useful by fixing different fields or adding missing data, such as expanding abbreviations or correcting misspelled words.

Presenting the Results

One of the main challenges for a data analyst is to work with the research lead to explore the data, figure out if anything stands out, and create insights and reports. Sometimes a wealth of data can lead to a poverty of insight. If a data analyst overloads the team, it can actually limit how everyone interprets the information.

To avoid this, a data analyst needs to work closely with the research lead to explain trade-offs in their reports. Often with statistics, it's what you don't see that's crucial to your understanding. Sometimes that's intentional; other times it's not.

Let's say the research lead wants to see a summary of all the men and women who shop on the running shoe web site and show a breakdown by age groups. As the data analyst, you might ask about the age brackets. Do you create a bracket for every five years or every ten years? If you create one for every five years, you have 18 or 19 brackets that are skewed in the middle because there are probably fewer runners younger than 18 or older than 90.

The report may also be difficult to read. You wouldn't have the type of granularity that you might need for people between the ages of 20 and 40. There could be a big difference between 35- and 40-year olds. If you break out that age bracket, you might be misrepresenting the data. It might make it look like there are fewer people between the ages of 30 and 40. That's why it's important for the data analyst to be transparent in how they present the data. There are decisions here that impact the story. The whole team should work to be up front about those decisions and communicate an accurate story, as shown in Figure 7-1.

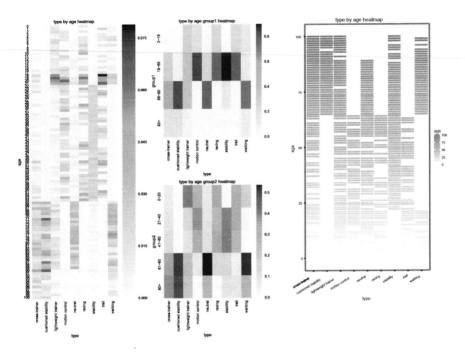

Figure 7-1. Heatmaps

▨ **Note**   Looking at the heatmap at the most granular level (each age as a cell) and at heatmaps of different age groupings: The left one shows that lightweight trainers are the most popular for ages 18-21; motion control, racing, stability, and trail are more popular for ages 21-65. However, because of the different age cutoffs, shown on the right, this data might carry a very different message. In the upper right, it seems that lightweight trainers are equally popular in all three age buckets (0-18, 19-65, 66-90), but if you look at the bottom right and the original heatmap, it's clearly more popular among only 18-21. This is because the first bucket on the upper right divides the most popular range right in half. See how to create these charts at http://ds.tips/m2zAm.

All of this decision-making would normally be at the data analyst's discretion. However, working with the research lead to make these decisions helps the whole team better understand the data.

Staying on Track with a Project Manager

One of the hardest parts of working on a data science team is making your insights actionable. The work should start to feel like an ongoing science experiment. You get a little bit of data. The research lead drives interesting questions. The questions spur more reports, which usually leads to more questions.

In this whirl of exploration, the team needs to focus on the purpose of gathering the data. There also needs to be a mechanism for explaining the insights to the rest of the organization. Unfortunately, this means many meetings, which can be a real time drain. This is especially true when the work is creative.

I once worked for an organization that had a very creative data science team. They were coming up with new and interesting ways of using the company's vast credit card data. For the first few months, the data science team was mostly left alone to explore the data. Eventually, the team produced some really interesting insights. As their insights became more interesting, the rest of the organization became more curious. Departments started calling on team members to give presentations. These meetings made the other teams even more interested. This led to more meetings. After a few months, some people on the data science team were in meetings for up to 20 hours a week. They went from data science practitioners to presenters.

The absence of these key people from the data science team slowed down the rest of the team to the point where they were producing fewer insights. They spent much less time looking at their data. The same departments who were requesting the meetings started asking why the data science team wasn't finding as many insights.

It's a paradox in many organizations. The more interesting your work is to the rest of the organization, the more time you spend talking about your work in meetings, which means less time actually working.

The best way to break this cycle is by having an effective project manager. Project managers are very good at protecting the data science team from getting off track. They can do this by representing the team at meetings.

A project manager on a data science team is responsible for three main areas: democratizing the data (making it widely available), sharing the results, and enforcing organizational learning. A project manager acts as both a shield and a bulldozer; a shield to protect the team from too many meetings, which will help keep the team working, and a bulldozer to break down barriers and get access to hidden data.

Democratizing the Data

When you work in a large organization, it becomes increasingly difficult to get access to all the data. A project manager can help break through these silos so the whole organization can benefit. For example, say your data science team wants access to data that is sitting on some dusty server in the corner of the data center. When the team tries to access it, they find out that it's off-limits and only available to a certain department. Your project manager is responsible for trying to convince that department to give the data science team access to that data.

Sharing the Results

Project managers also work to distribute information. They are the ones who go to the meetings and present the team's results. If someone from another department wants access to the team's data, the project manager works to give them access.

Democratizing data and distributing results are closely related. It's a two-way street: one way gets access to organizational data and the other way grants access to the team's insights to the rest of the organization. Each of these has its own set of meetings and challenges.

Enforcing Organizational Learning

The final area for the project manager is enforcing learning—taking the insights and making them actionable. At the end of the day, the team will still be evaluated based on what the organization learns. Someone needs to follow through and turn the insights into products or changes.

Imagine if that organization with the credit card data had found something interesting, such as a slight uptick in transactional data around lunchtime. The project manager would turn that insight into something actionable. He or she may also work with the infrastructure group to scale up their technology during those times, or work with the marketing group to create lunchtime promotions. Enforcing this learning takes a lot of time and a lot of meetings. A good project manager keeps the team on track by taking on this responsibility.

Working As a Team

Once you have your team in place, the entire team works together to see if they can create an interesting data model that show trends in the data. Maybe you'll see a strong correlation between a few different items, and then work together to guess at what causes the correlation.

Let's say that your team works for an online magazine. At the top of each story, there's a link that allows readers to share the article with their social network. The data analyst on your team ranks the stories that have been shared the most by readers and prepares a report, shown in Figure 7-2, for the team so you can discuss the findings. The research lead, in turn, might ask, "What makes these articles so popular? Are they the most widely read or the most widely shared? Are there certain topics that make an article more likely to be shared? Are there key phrases?"

Figure 7-2. Counts by channel and title along with totals

▓ **Note**   This data consists of randomly generated article titles and share count by channels. From the chart, you can tell even though they are all top stories in terms of share count, the channels can be very different from one to another. See how to create these charts at http://ds.tips/spu3E.

Your team works together to create a model that shows a correlation between certain topics and the likelihood of that topic being shared. Maybe the team creates their own topic identification. The research lead is critical here because she is the one who knows the most about the business. She might know enough about the readers to guess at certain topic categories. She might be the best resource for coming up with keywords, such as "sneak peek," "blurry photo," "quirky humor," or "rumors," which suggests technology rumors. (Someone from a publishing background would have a lot of useful suggestions.)

The team could even create the first version of a program that tries to sort articles into these categories based on keywords. The data analyst creates the reports and develops the application that identifies the topic of the story. Then, he creates a data model that accurately sorts the most-shared articles. (This person may not know as much about publishing, but he knows how to work the data.) Finally, the team uses predictive analytics to apply that model to the future. You now have an application that can accurately predict when a new article is likely to be shared by many of the readers.

This is where the project manager steps in, takes this new insight, and makes it actionable. She communicates the results to other teams and works with management to improve the organization. She may even work with the marketing department to put the most profitable advertising on the articles that are more likely to be highly shared by readers.

Part of the data science mindset is acknowledging that, to encourage innovation, you need the research lead and data analysts to work together to promote new ideas. The research lead has to drive interesting questions. They also have to foster diversity of opinion. They may want to bring in people from other parts of the organization.

If you run your team with just data scientists, you're likely to lack a significant diversity of opinion. There will be too many similarities between their training and backgrounds. They'll be more likely to quickly come to a consensus and sing in a monotone chorus.

I once worked with a graduate school that was trying to increase their graduation rate by looking at past data. It turned out that the best idea came from a project manager who was also an avid scuba diver. He looked at the demographic data and suggested that a buddy system might increase the number of students who went through the whole program. That was common practice in scuba training. No one could have planned his insight. It just came from his life experience. Figure 7-3 shows an example of the data you can collect about the buddy system.

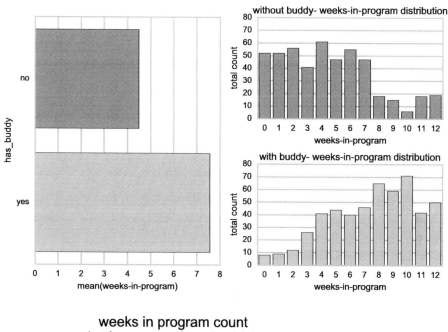

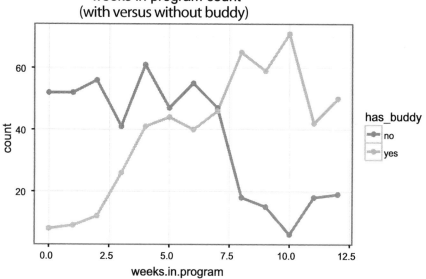

Figure 7-3. Buddy system data

> ▧ **Note** Looking at the left, if a student participates in their buddy program, on average, that person will stay three weeks longer in the program than those who did not have a buddy. Those who have a buddy are more likely to stick around after eight weeks than those who did not. See how to create this chart at `http://ds.tips/naF7u`.

This form of creative discovery is much more common than most organizations realize. In fact, a report from the patent office suggested almost half of all discoveries are the result of simple serendipity. The team was looking to solve a problem, and someone's insight or experience led them in an entirely new direction.

> ▧ **Note** We cover this working relationship in more detail in Chapter 8.

Summary

In this chapter, you learned that one of the most effective team models in data science includes the following roles: research lead, data analyst, and project manager. The research lead has three areas of responsibility: find assumptions, drive questions, and know the business. The data analyst prepares the data, selects the tools, and presents the results. Finally, the project manager gathers the data you need from various departments, then shares the team's results and enforces organizational learning. In Chapter 8, you'll find out how this team will start to work together.

Starting the Work

Now that you have your team together, you need to understand how to start to work. First, you'll need to explore the areas of responsibility for each team member and indicate where those areas of responsibility overlap. The next thing you need to think about is how to present your data. Data visualization is a topic that is covered in entire books. I will touch on this topic briefly and cover the two different types of reports your data science team will mainly focus on: internal and external.

After you know how you're going to present your reports, I will move on to explaining data silos and how they affect your team's ability to collect data. Then, I'll give you some tips on how to democratize your data. Finally, I will cover the importance of providing insight into the data science team and getting executive buy-in.

Defining Areas of Responsibility

A data science mindset is a big change from how most organizations operate. Even organizations that call themselves data-driven don't often use their data to create new insights. Instead, they use data the same way that a drunken person uses a lamppost.[1] They think of it as support and not illumination.

---

[1] This saying is often attributed to Scottish novelist and folklorist Andrew Lang.

© Doug Rose 2016
D. Rose, *Data Science*, DOI 10.1007/978-1-4842-2253-9_8

This is a real challenge for your data science team. Your organization could think it's data-driven, but they're actually just using data to reinforce what they already know. Anything that contradicts this knowledge is seen as bad data.

Your data science team needs to make sure to use data for discovery, which keeps the team from falling into the trap of just using data to support what's already known. In fact, a major benefit of data science is questioning established knowledge. It's like that old Mark Twain quote: "What gets us into trouble is not what we don't know. It's what we know for sure that just ain't so."

If your organization is relying on knowledge that's not backed up by the data, you're likely to run into trouble. Often, this shared knowledge is right, but when it's wrong, it can have lasting consequences. If your data science team stays true to its three areas of responsibilities, it can be a real benefit to your organization. The three areas are research, question, and implementation, as discussed in the following sections.

So far you've seen three common roles on a data science team: The research lead who drives interesting questions, the data analysts who work with the research lead to come up with interesting reports and insights, and the project manager who makes those insights actionable and available to the rest of the organization.

Now is the time to take these roles and place them into larger areas of responsibility so you can see how the team comes together. Think of your team as having different overlapping areas of responsibility, as shown in Figure 8-1.

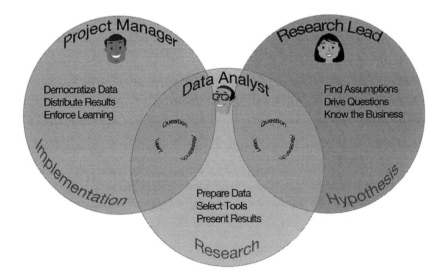

Figure 8-1. The overlap between different roles

Researching

Let's start our Venn diagram with the data analyst. The primary area of responsibility for the data analyst is researching, which is a key part of science and data science. The data analyst works with the research lead to come up with interesting questions, research these questions, and represent the results with a creative report or chart.

That data analyst is the foundation for the team. He works with both the project manager and research lead. They just work in different ways. Think of him as having an independent two-way relationship with both of them. He gets inputs from the research lead in the form of interesting questions, and then he outputs the results and insights to the project manager so she can enforce learning.

Questioning

Now let's look at the next circle in Figure 8-1. On the right, there's the circle for the research lead. Their area of responsibility is questioning. If you think about this in terms of a scientific method, this person is the one who creates an interesting hypothesis.

The research lead creates a cycle between herself and the data analyst. She's asking questions and getting feedback. It's not as simple as sending an e-mail to the data analyst that says, "What do you think?" It's a collaborative process. The research lead asks questions and the data analyst provides feedback to those questions based on the available data. These two circles overlap. It's a direct relationship between questions and research.

Implementation

The final circle in the Figure 8-1 is for the project manager. His area of responsibility is implementation. He needs to make sure that the team takes the data and uses it for something actionable. He makes sure the team distributes their insights to the rest of the organization.

It is not an easy challenge to take an exploratory process and apply it to organizational knowledge. On the data science team, you often don't know the avenues of your most actionable insights. The team will go through 50 dead ends before the team finds one interesting path.

Still, it's important to consider what these insights might look like when they're finally implemented.

Each of these areas of responsibility is part of a map of what the team needs to accomplish. It helps reinforce the idea that your data science team is about exploration and discovery. The team needs to follow the data even if it contradicts established knowledge.

Presenting Reports

Data visualization is one of most interesting areas of data science. It is how you display your data using graphics and imagery. Simple graphics are one of the best ways to communicate complex ideas. It can be a real challenge to balance complex ideas against simple design.

Most reports tip too far in one direction or the other. There are many beautiful charts that don't really communicate much information, and there are informative charts that are ugly and difficult to understand. Good visualization is a core responsibility of the data analyst. That being said, the analyst should work closely with the data science team. If you have to explain the chart to the research lead, it's probably too complex for everyone else. The team is a good testing ground to help you make your charts more beautiful and readable.

There are many good resources on data visualization. One of the oldest and most well respected is *The Visual Display of Quantitative Information, 2nd Edition* by Edward R. Tufte.[2] In this book, Professor Tufte introduces the idea of the data-ink ratio. He states that you should push your visualizations to communicate the maximum amount of data with the minimum amount of ink. He uses the term **chartjunk** for useless visuals such as 3-D shadows or gradient effects.

Prof. Tufte helped establish and set the direction of the field of data visualization. He helped establish a modern approach to presenting data.

For another approach, check out *Storytelling with Data: A Data Visualization Guide for Business Professionals* by Cole Nussbaumer Knaflic.[3] In her book, she lays out six key lessons. She starts off by saying a data analyst should understand their audience and the context; choose an appropriate display, eliminate clutter, and focus attention; and think like a designer to tell a compelling story. These two books will give you a good sense of what it means to create nice data visualizations.

Note   There's typically nothing in the training of a data analyst that prepares them to create good visualizations. Most graduate programs are still very much rooted in math and statistics. Good data visualization relies on aesthetic and design. It's a learned skill and may not come easy.

---

[2]Tufte, Edward R., and P. R. Graves-Morris. *The visual display of quantitative information.* Vol. 2, no. 9. Cheshire, CT: Graphics press, 1983.
[3]Knaflic, Cole Nussbaumer. *Storytelling with Data: A Data Visualization Guide for Business Professionals.* John Wiley & Sons, 2015.

Remember that your team works together to explore the data, which means that the majority of the first round of reports that you design will be for each other. The research lead drives interesting questions, the data analyst creates a quick and dirty report to explore possible answers, and then the team might come up with a whole series of new questions. This means that most of your initial data visualizations will be quick exchanges—more like visual chitchat than full data reports.

On the other hand, there are the final data visualizations that you need to communicate with the rest of the organization. These visualizations will be polished, simpler, and easier to understand.

Think of your first round of visualizations as data whiteboards, like the whiteboards on the walls of most organizations. They're used primarily for quick visualizations to help with small meetings and discussions. Your first report should have that level of readability. It might be a quick and simple cluster diagram. Maybe it'll just be a simple linear regression.

No one would ever take a picture of the scribbles on a whiteboard included in an executive report. It's just for the team. To take the data representation from the whiteboard to the executive report, you have to add clarity and polish.

If you're a data analyst, remember to take advantage of the feedback you get from the rest of the team. Data visualization is like any design. You don't need to be an expert to have an opinion. Your team will be the best way to figure out if your chart is readable and understandable. Also remember that your best charts will be the product of an emergent design. Start with simple reports and improve them over time to make them simpler and more readable. Each new version should clarify the information you have. You can also improve your report by adding information without increasing its complexity.

You'll end up with much more beautiful reports if you go through several improved revisions. The rest of the team will be the best resource to help you get there.

Understanding Silos and Freeing Your Data

One of the biggest challenges for a data science team is getting access to all of the organization's data. This is one of the core responsibilities of your project manager. She'll work hard to get access to the dusty old database servers in every corner of your organization.

It's not uncommon for large organizations to have a different database for each department and not share information. Each department will have its own data analysts, manager, and database engineer. The people in that department will know about their data challenges, but not know much about other parts of the organization. This is typically called a **data silo**. The term silo is borrowed from agriculture. Farmers typically store seasonal grains in a tall, hollow building. If you've ever seen one, you know a silo is a tall, standalone building. Each silo stores its own grain, and there's usually no connection with other silos. It's the same with organizational data. Each department stores bits of data, and it isn't mixed with other silos.

Data silos exist for a good reason. Each department might have its own regulatory challenges or security measures. One database might have passwords and another might just be a simple list of products. Chances are the password database is encrypted and secure, while the product database is open and available.

The problem with silos is that they make it very difficult for your organization to have a holistic view of your data. I once worked for an organization that was trying to convert all of its data into one centralized cluster. They felt that they weren't getting enough insight from their data. The organization had just gone through a data governance transformation and wanted to govern how the data was controlled.

When they finally got into their data, they realized how much was locked away in silos that no one knew about. Over many years, each department created its own processes, schemas, and security procedures. The organization wanted to get value from this data, but it was on different servers all over the company. Asking everyone to share their data was a little bit like asking them to share their toothbrushes. The project manager was in many meetings with very heated discussions.

Unfortunately, breaking down these silos is necessary to become a more data-driven organization. You will want several data science teams working on one centralized data cluster. You will also want people outside the data science team to create rudimentary reports and charts.

Remember that data is the key material for finding insights and creating new organizational knowledge. To be data-driven, you need to have free access to data.

Now that you understand silos, how do you **democratize** data so it's accessible by as many people as possible? You need to make sure data is no longer a protected resource that's passed between executives and data analysts. Instead, it needs to be a company-wide resource that is manipulated and understood by everyone.

If you're a project manager on a data science team, try to keep a few things in mind:

- Don't underestimate the difficulty of getting access to data silos. It takes a long time, and you want to get started before the team actually needs their data.

- It takes an organizational change to create a centralized data cluster. If you don't have executive buy-in, you probably won't make much progress. You probably need to sell each department on the idea of a centralized data cluster. Most departments will not share your view. They figure if it's not broken, why should they spend time trying to fix it? You might have to entice them by explaining that they will be able to create more complex reports or use newer visualization tools.

- You need to provide access to your team's reports. You might have an easier time breaking down silos if you can show the value of company-wide reporting and insights. After awhile, it will be easier to make an argument that shared data is like any other shared resource. The whole will be better than the sum of all its parts.

Do your best to protect the data science team from the data silo meetings. You want them focusing on exploration and discovery. You focus on increasing their usage and access.

Creating a Cycle of Insight in a Data Science Team

Many organizations focus on monitoring each team's milestones. The managers focus on their compliance, and much of their effort is dedicated to planning. They have quarterly budgets and monitor them closely. They look for cost or schedule variances. If they see a change, they quickly chase it down, and then report it to executives. These types of organizations are structured for monitoring and compliance. If you are in this type of organization, think about your meetings. Chances are, you're doing things like planning or presenting a plan, coordinating with another team, asking for a budget increase, or asking for a schedule extension because you're running behind.

Working this way is not well suited for data science teams. Remember your team's work is exploratory. Its members come up with questions, create theories, and then run experiments.

There are certainly companies that are used to working with scientists, such as pharmaceutical or high-tech companies. These types of companies have been running experiments for years. But for most companies, exploratory work is a new concept. It won't seem natural to have a data science team that creates new knowledge. In these companies, you have to be especially careful about how the team works. There will be institutional pressure to separate the business from technology as well as a strong push to make sure that a compliance manager runs the team. This is usually a project manager or a director. Putting these structures in place can slow down the pace of discovery.

I once worked for an organization that wouldn't let a research lead work closely with the data team. They felt that it was the data analysts' job to come up with the best insights, and then a business manager would just see the ideas in monthly presentations.

The business manager had her own budget, separate from the data team's. The business manager wasn't interested in finding insights. She just made sure that her team stayed within their budget. Having a full-time research lead was not something that was within their budget. The data science team stopped exploring before it even started.

I saw a different company try to use the project manager to monitor the data science team's milestones. He tried to create different ways to measure the team's progress. He created tasks for developing questions, and then measured how well the team completed these tasks. It didn't work well because most questions just lead to more questions. The project manager wasn't happy when the team's milestones kept slipping. His incentive was to "complete" the questions as quickly as possible, which is the opposite of what you want in a data science team.

When you're on a data science team, try to be aware of these institutional pressures. Most organizations have a tough time accepting a data science team that can't be easily measured or controlled. It's difficult to set goals or create a return on investment.

Work hard to make sure your team doesn't get pulled into these compliant structures. Instead, make sure the team creates a feedback loop. Everyone should work together to question, research, and learn.

Each person on a data science team has his or her own focus area, but they're still working together in tight feedback loops. For example, the project manager participates when the research lead and data analyst explore the data, and the data analyst might give the project manager some good suggestions for getting access to another team's data. The team will always do better when everyone is participating. When you're exploring, the more eyes you have on a problem, the more likely it is you will have great insights.

Also, make sure that your data science team has executive-level support. Without it, you'll almost certainly be pulled back into common control strategies. It won't be easy, but if you have executive-level support and can create tight feedback loops within the team, you will have a much easier time making discoveries. Don't get frustrated if it takes a long time to make these organizational changes. The first step is to understand the purpose of data science and how it fits outside these long-serving organizational structures.

Summary

In this chapter, you explored the areas of responsibility for each team member. Then you saw where each of those areas of responsibility overlap. You also learned a little bit about how to present your data (data visualization) and the difference between internal and external reports. Next, you found out about data silos and how they affect your team's ability to collect data. Then, you received some tips on how to democratize your data. Finally, you learned the importance of providing transparency into the data science team and how that helps with executive buy-in. In Chapter 9, you'll find out how to get your team to *think* like a data science team.

Thinking Like a Data Science Team

Now that you know how to structure your team and have divided the areas of responsibility, how do you ensure that your data science team thinks like a team? In this chapter, I help you out by examining some common ways to keep your team on track. First, I cover how to keep from reporting without reasoning by asking interesting questions. Next, I explore how you can find the right mindset for your team as a whole. And finally, you learn about making sense of the data and get some tips on how to find your way out of team freezes.

Avoiding Reporting Without Reasoning

If you're not familiar with statistics, a great place to start is *Naked Statistics: Stripping the Dread from the Data* by Charles Wheelan,[1] a former public policy professor from the University of Chicago. It's a fun read and a great introduction to statistical analysis. In the book, he goes over the dangers of using reports to draw sloppy conclusions.

---

[1]Wheelan, Charles. *Naked Statistics: Stripping the Dread from the Data.* WW Norton & Company, 2013.

© Doug Rose 2016
D. Rose, *Data Science*, DOI 10.1007/978-1-4842-2253-9_9

If you know what to look for, you can see sloppy conclusions everywhere. One place you'll see this is on Internet news sites. Professor Wheelan imagines a news site with an attention-grabbing headline: "People Who Take Short Breaks at Work Are Far More Likely to Die of Cancer." Sounds pretty scary. According to this study of 36,000 workers, those who reported taking several ten-minute breaks each day were 41% more likely to develop cancer over the next five years. Those who didn't take these breaks were much healthier, as shown in Figure 9-1.

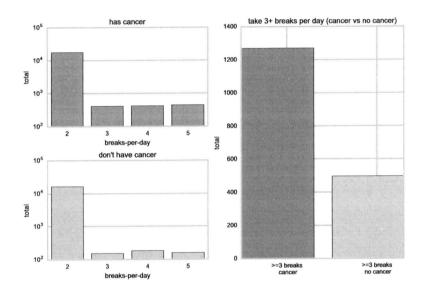

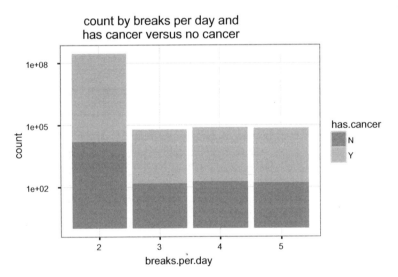

Figure 9-1. People who take short breaks are far more likely to develop cancer

If we only look at those that take three or more breaks per day, they are indeed about 43% more likely to have cancer. However, 95.1% of people only take two breaks per day. See how to create this chart at http://ds.tips/Sp4ye

A sloppy data science team might report this to the rest of the organization, and the company might take outlandish steps, such as locking the doors to ensure everyone's safety. This might seem like an extreme example, but it's much more common than you'd think.

A responsible data science team would never draw conclusions from this type of report. Instead, the research lead would use it to drive interesting questions. Why would getting up from your desk be so deadly? Who are these people who are taking several 10-minute breaks? Why are the people who aren't taking these breaks that much safer?

The research lead might have to research these questions by talking to people familiar with these workers. Maybe someone will recognize this behavior.

If you have worked in an office, you probably recognize that anyone who gets up several times a day for 10 minutes is probably going outside to smoke a cigarette. Remember that correlation does not imply causation. In this case, the problem was that these workers were frequent smokers. The connection with them getting up every 10 minutes was just incidental. It was the smoking that was the real danger.

While working on your data science team, remember that the best defense against sloppy reporting is working with the research lead to ask interesting questions. Remember that conclusions are easy. It's the exploration and reasoning that is difficult. These will be the source of your greatest insights.

There should be a healthy tension between your research lead and your data analyst. The data analyst will be looking for data to draw conclusions. The research lead will always have further questions. The data analyst will present reports and the research lead will test the weaknesses of those reports. In the end, this should help create stronger analysis.

Sloppy reasoning is a much larger problem than it may seem. The previous example was just a fake Internet headline, but sometimes the reality can be dangerous. In the 1990s, doctors observed that women who took estrogen supplements were one-third less likely to have heart attacks. A massive study of 122,000 women showed that there was a negative correlation between estrogen supplements and heart attacks. Many doctors concluded that estrogen had a protective effect that might be beneficial to female health.[2]

By 2001, 15 million women were taking estrogen supplements, commonly known as hormone replacement therapy. Then a strange thing happened. Doctors started to notice that women who were on hormone replacement therapy were much more likely to have a stroke, heart disease, and breast cancer.

Years later, doctors began to scrutinize the findings. Many of them concluded that in the initial study, the women who appeared healthier had external factors—they were wealthier, more health-conscious, and more likely to have excellent medical care. The estrogen wasn't the likely cause of their good health. This data is still debated today. One thing that is accepted, however, is that this sloppy analysis contributed to the deaths of thousands of women.

The best defense against sloppy reasoning is the creative tension between the research lead and the data analyst. If you're not asking the right questions, you're much more likely to draw the wrong conclusions.

Having the Right Mindset

You've seen the dangers of sloppy reporting. Now let's think a little bit about getting into the right mindset. Many organizations think that data science is just an analytical skill, that there are bunch of analysts in a room who simply report back with their numbers. These numbers are seen as the truth because the numbers don't lie, but numbers do lie. You saw examples of statistics telling the wrong story in Chapter 4, where two politicians told different stories based on the same numbers. Studies can reach false conclusions. People can misinterpret the data. Their story can be incomplete.

You've already seen that asking questions is the best defense against sloppy conclusions. So how do you come up with better questions? To get there, you need to get into the right mindset. The good news is that over the last few years, a lot of work has been done in this area. Many different fields have come together to better see how people think. Computer engineers apply these to machine learning and artificial intelligence.

---

[2]Shlipak, Michael G., Joel A. Simon, Eric Vittinghoff, Feng Lin, Elizabeth Barrett-Connor, Robert H. Knopp, Robert I. Levy, and Stephen B. Hulley. "Estrogen and progestin, lipoprotein (a), and the risk of recurrent coronary heart disease events after menopause." *Jama* 283, no. 14 (2000): 1845-1852.

The research has found that analytical thinking won't necessarily serve data science well. The reports and data are only the first step. The next step is conceptual thinking—the ability to look at data and combine it with your own intuition. Conceptual thinking will help the team identify areas to focus their questions on.

There is an interesting book on this topic titled *A Whole New Mind: Why Right-Brainers Will Rule the Future,* by Daniel Pink.[3] In the book, he argues that we're near the end of the information age, and that focusing on just numbers and reports isn't that valuable. The real value will come from the knowledge that we create. He calls this the conceptual age.

Data science is at the very edge of this conceptual age. At some point, the analytical tools will be simple enough so that many more people will have access to the data. Soon, users will get access to data science tools that are as accessible as WordPress or Microsoft's LinkedIn. This will give more people access to interesting information.

For now, the data science team is responsible for both the data and its underlying concepts, which in many ways is much more difficult. The team will have to use their analytical skills as a starting point, and then use an entirely new set of conceptual skills. In Daniel Pink's book, he goes through several of these skills. He calls them the "senses" of the conceptual age. I've adapted these senses into three team values. These values should help your team think about the data at a conceptual level.

Storytelling over Reporting

The first value is **storytelling over reporting**. Your data science team should strive to deliver interesting stories about the data. You should tell a compelling narrative. More often than not, your data will be interpreted, which means there can be more than one story. It's easier to come up with questions if you think about an interesting story.

Think about the data as characters in a play. Ask why they are doing one thing instead of the other, and then ask questions about their behavior.

Symphony over Detail

The second value is **symphony over detail**. One of the best ways to participate in the information age was to specialize. You see this in how companies hire— for example, database engineers often specialize in just a few platforms. You'll want to move away from this specialization on your data science team.

---

[3]Pink, Daniel H. *A Whole New Mind: Why Right-Brainers Will Rule the Future.* Penguin, 2006.

You want the team to create a symphony by putting together several different stories and creating a bigger picture.

You've already seen this in a couple of examples: seeing the big picture with smokers leaving their desk and understanding why customers purchase running shoes at the beginning of the year. These stories require you to bring in many different types of data to come up with a greater understanding.

Empathy over Certainty

The final value is **empathy over certainty**. Knowing what motivates people is one of the best ways to come up with questions about your data. Your data science team will want to understand what your customer thinks about and what is important to them. Remember that data science can analyze the actions of millions of people. If your team can understand what motivates them, they can ask more interesting questions.

It's important to remember that your data science team will have to use an entirely new set of skills to succeed. To ask good questions, you have to think conceptually. Try to use these team values as a way to remind yourself that data science is not about simple reporting. Your team needs to use their conceptual skills to ask good questions and create organizational knowledge.

Going Deep into "Sense-Making"

As you've seen, it's very important for the team to have the right mindset. Your team should think conceptually about the data. That's not an easy task. Remember that most data analysts come from math or statistics, which are fields that tend to rely on structured metrics.

Conceptual thinking often requires a storytelling style, which is more creative and artistic. A lot of data analysts need to lean on their abilities and not their training. Once everyone is in the right mindset, they can start to focus on sense-making. Sense-making is a key part of data science.

We all do sense-making in one way or another. When you come back from a vacation, you might be overwhelmed by the e-mail in your inbox. You might decide to create subfolders labeled "important," "old," or "announcements," or you might decide to sort all of your e-mail by sender.

Each person may approach the data differently. As a team, it's even more difficult to have a shared sense of the data. Each person has his or her own technique for sense-making. These views may not overlap. What makes sense to one person might look like a waste of time to the other.

Sense-making can be an enormous challenge for data science teams. These teams will attempt to make sense of very large data sets, and the information can be overwhelming. This might lead to **team freeze**, which is when a team has so much data that they're not sure where to start. If you're outside the team, it might be difficult to see when the team is freezing.

I once worked for a company that was trying to make sense of an enormous new data set. The company collected modest amounts of data and then purchased large data sets from an outside company. The hope was that they could link their existing data to these larger data sets to better understand their customer. With the help of the other company, they were able to quickly get these new data sets into their cluster. The problem was that the data science team couldn't figure out where to start. They struggled with sense-making. They froze and just produced reports that showed how much data they had.

They were frozen for months. Whenever they had a business meeting, the team just showed nicer reports of the data about the data. They downloaded expensive visualization tools and produced terrific charts. But in the end, they weren't asking any interesting questions.

If you're on a data science team, try to look for signs that you might be struggling with sense-making. If everyone is focused on the tools, it might be a sign that your team is overwhelmed by the data. Watch out for empty demonstrations that take the place of data science.

One way to get past data freezes is to have more fun with the data. You can come up with ridiculous questions that might not seem to have value. Try to come up with questions to see if someone's a dog or a cat person. Maybe try to guess someone's height by just looking at his or her purchase history. Remember that questions often lead to more questions. Just getting your hands on the data and playing with it is often enough to get started. A lot of your team's discoveries will come from serendipity, which means the more you play with the data, the more likely you'll find something interesting.

Don't be afraid of toying around (but don't call it that to people outside the team). Inside the team, just bounce around questions until something sticks. There's often a thin line between experiments and playing games. The more you're in the data, the more you get a feel for some interesting questions.

Try to understand that each member of your team might approach the data in a different way. This is a key part of sense-making. You should call out this challenge and look for a way to have a shared sense of the data. It's also important to recognize when your team is freezing. Many data science teams will get stuck looking at the tools when creating reports. Finally, remember to have fun. You may get some of your best perspectives by playing around with the data.

Summary

In this chapter, you learned some common ways to keep your team on track. These include how to keep them from reporting without reasoning by making sure they ask interesting questions. You also found out how to have the right data science mindset. The team needs to make sense of the data by thinking conceptually and learning to tell stories. In Chapter 10, you'll find out how to avoid pitfalls when building your data science team.

Avoiding Pitfalls in Building Your Data Science Team

In this chapter, we cover the two of the main pitfalls that affect data science teams. First, if a team reaches a consensus too quickly, it stifles discovery and is a sign that the team has blind spots and is prone to groupthink.

Finally, how do you know when your team is spending too much time on the wrong question or asking questions about the wrong things? This is called wandering and we provide you some tips on how to avoid that as well.

Steering Clear of Consensus

In most organizations, people naturally try to reach consensus. Different organizations call it different things. Some encourage everyone to "go along to get along." Others use words like "socialization." Data science is very different.

© Doug Rose 2016
D. Rose, *Data Science*, DOI 10.1007/978-1-4842-2253-9_10

Consensus can be a big problem. You want your team to explore new ideas. If everyone comes to a consensus too quickly, that could mean that everyone shares a common misunderstanding.

Remember that data science is about exploration. You're looking for knowledge and insights. There's no need to get everybody on the same page. In fact, you want everyone comfortable enough to argue about how to interpret the data. The data science team should be more like an awkward family dinner than a quiet bus ride. You want the team to talk, explore, and even annoy one another. This type of exchange is much more likely to uncover new ideas.

There are a few things you can do to keep your team from quickly reaching consensus:

- Be aware of the danger of consensus. Recognize that a quick agreement on a complex topic is often a sign of groupthink.

- Make sure that you keep your team small enough that everyone feels comfortable disagreeing. Your team should have your research lead, a few data analysts, and the project manager. Strive to keep your team at fewer than six people. Larger groups tend to crowd out quieter voices, which often have some of your best insights.

- Make sure that your research lead brings in people from outside the team.

Let's expand on this last point, because it's critical to staying away from a quick consensus. Let's say you're looking at an interesting question about your running shoe web site. You're trying to see if people who are in a romantic relationship are likely to talk their partner into running. The team is trying to figure out how to get at this question through data from the web site. The research lead might want to invite someone from sales to share anecdotal stories or include a few people they know who run with their girlfriend, boyfriend, husband, or wife to talk about how they got started running with their partner. These people might be able to add some insights that the team might not otherwise have. Just remember to keep the team small. So maybe just add one or two people per meeting.

Something else your team may try is to end every discussion by assuming that they're wrong. This could be as simple as asking, "What if we're wrong about this?" Your team should be able to answer this question. Maybe they'll realize they're wrong about this or they're wrong about many other things. If that's true, go through all those other things. Look out for the times when the team has no answer to this question. That is one of the warning signs that the team might be falling victim to groupthink.

Let's say the team decides to approach the question of romantic running partners. They decide that the best way to look at this data is to see if there are customers who share the same address, and then compare their orders to see if one customer started ordering before the other, as shown in Figure 10-1. But what if the team is wrong? What will the data look like? Maybe roommates are likely to run together. They could be students or trying to split rent. These may be weak arguments, but they should still be considered. It's the discussion about what it means to be wrong that's most valuable.

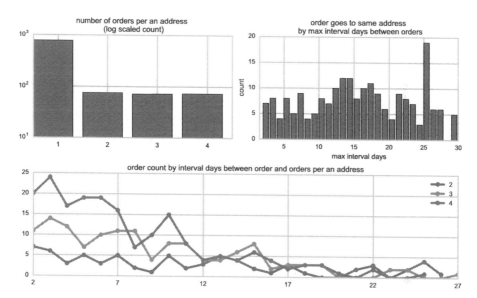

Figure 10-1. Orders from customers who share an address

Most addresses have only one order, and there's a maximum of four orders per address in a one-month window. Excluding those with just one order, on average, the max interval days between orders to one address is 16.5 days. If we compare the interval days order to order, most likely people will place another order within 15 days, and the more they order within a one-month timeframe, the more orders they will place within the first 7 days. See how to create this chart at http://ds.tips/s5Ere

Another thing to try if the team is reaching consensus too quickly is to find someone who is likely to disagree and be the devil's advocate. For example, for the running shoe web site, you might want to bring in someone who would never run with a romantic partner. Ask her why she feels this way. Maybe it's much more common for people to use that time to be alone. It could be that runners like running because it's a solitary sport. These are the kind of insights that can really add a lot to the team's discussion.

If you work for a large organization, you might be more inclined to reach a quick consensus than you realize. A typical project in a large organization requires a lot of coordination. For these projects, you need a consensus to get things done. A data science team has a different set of priorities. When you're looking for new knowledge and insight, reaching a consensus too quickly can be a real danger. If you use the techniques discussed in this section, you're more likely to keep your team exploring and open to new ideas.

Keeping the Team from Wandering

As we've seen, there's a danger of the team reaching a consensus too quickly. On the flip side, how do you know when your team is spending too much time on the wrong question? When should you abandon a question and start another? This is the opposite challenge. Instead of everyone agreeing, they continue to ask questions about the wrong things. The team ends up wandering instead of exploring.

In many ways, this is a much more difficult challenge than consensus or groupthink. You want your team to feel the freedom to wander. At the same time, the business needs to gain valuable insights. You don't want to stifle innovation by focusing on delivery. Yet the team has to deliver.

I once worked on a project for a large home improvement retailer. They were trying to determine whether customers were regular people or professional remodelers. They tried to create a predictive data model based on what customers purchased.

The research lead asked some very interesting questions. What items are professionals more likely to purchase? Are there times when a professional is more likely to shop? Maybe they shop early in the morning before they have to report to a construction site. Are professionals more likely to make large purchases?

All these questions were very interesting. The data science team brought in a few outsiders so they could gain their perspectives. They did a great job avoiding consensus. There were many different questions. There were also a few different ways to get at the data.

One challenge that the team had was that the retailer mistakenly thought if they had more data analysts they'd be more productive, and made the team much larger. There was one research lead, a project manager and four data analysts.

The research lead would drive interesting questions, and then the team of data analysts would produce several different reports. The problem was that each of these reports was very narrowly defined. This will sometimes happen with data analysts. Most analysts come from the structured world of statistics, mathematics, or software development. They've been trained so that when they see complexity they tend to break it down into narrow metrics. So instead of being overwhelmed with the data, they become underwhelmed with small details.

That was exactly what happened with this team. They produced dozens of reports every few weeks with small, uninteresting conclusions, as shown in Figure 10-2. They found that people who bought paint are more likely to buy in the morning, people who made large purchases were more likely to buy an appliance and customers who bought carpet were more likely to buy it on Friday.

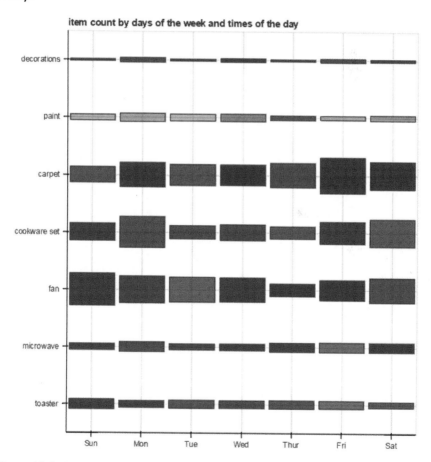

Figure 10-2. Item counts by days of the week and times of the day

---

▨ **Note** The larger the square, the more money was spent on a given day of the week on that item. The darker color represents sales made later in the day; the lighter color represents sales made earlier in the day. For paints and decorations, the total values are quite small compared to carpet, cookware, and fans. Paints are mostly bought in the morning. Carpet has a slightly higher sales value on Friday and cookware has slightly higher sales on Monday. See how to create this chart at http://ds.tips/y2wRe.

Each of these little data points was fine, but there was no attempt by the data analysts to create a story. What are the motivations and challenges of a professional customer?

This is the problem with looking through the keyhole of a glass door. It happens sometimes when a team has too many data analysts. They focus on dozens of small data points instead of trying to wrestle with a larger truth. When this isn't addressed, the team can go on wandering into the data and lose interest in asking big questions.

It's a key responsibility of the research lead to make sure this doesn't happen by driving the team to ask interesting questions and deliver valuable insights. The research lead represents the business interests on the data science team.

You want the team to feel free to explore, but at the same time, you want to keep them from wandering. If the team isn't delivering insights, they won't be very valuable to the business. That's one of the reasons you want to keep the team small and equally represented—one or two data analysts, a research lead, and a project manager. If your team is too heavily weighted with data analysts, you're in danger of gathering data without insights.

Small teams can ask big questions. Also, remember that you own the knowledge and insights that come from the team. If you think the team is not thinking big enough, you have to challenge them to go after larger stories.

Summary

In this chapter, you learned some tips on how to avoid two common pitfalls in data science teams. One was reaching consensus too quickly and the other was wandering. If a team reaches a consensus too quickly, it stifles discovery and is a sign that the team has blind spots. If your team is wandering, they're spending too much time on one question. They also may be asking questions about the wrong thing. In Part IV, you'll learn how your data science team should focus on delivering value, starting with Chapter 11, where you'll see the differences between how a data science team behaves and how it's different from the way most organizations operate.

Delivering in Data Science Sprints

You've seen what data science is and how to form your data science team. Now you'll find out how to start the work. Data science requires a very unique mindset. You'll see how companies typically work and compare that to how a data science team works. Then you'll find out how to use a data science life cycle and deliver real business value in team "sprints."

A New Way of Working

The "science" in data science refers to using the scientific method. This scientific method is a loop of discovery. Your team will ask interesting questions, and then you'll research those questions. Next, you'll use the research to come up with new insights. Your team needs to take an empirical approach to the work. Instead of planning, they'll need to adapt. Instead of relying on answers, they'll look for interesting questions.

That's much different from how most organizations operate. Most organizations rely on planning, objectives, and concrete deliverables. Often in these cases:

1. Portfolios are broken up into smaller, actionable projects.

2. Each project may have several teams.

3. A central Project Management Office (PMO) tracks the project's success.

4. The PMO makes sure the teams are on track to finish the deliverable.

Your data science team, on the other hand, needs adaptability, exploration, and creativity to help them question, explore, and react.. That's a pretty big disconnect. Most organizations still crave planning, objectives, and certainty.

© Doug Rose 2016
D. Rose, *Data Science*, DOI 10.1007/978-1-4842-2253-9_11

For your data science team to be successful, you'll have to rewire how the organization thinks about your work. You have to shake loose the notion of planning and delivering and replace it with the notion of exploring and discovering. This change won't be easy in most organizations.

The first thing you have to do is communicate what makes data science different. Start by comparing data science to a typical project. Show how traditional notions of planning and delivering don't work well for a data science team. Redefine the success criteria by explaining that you can't focus on a big bang deliverable at the end of the project. Instead, you have to show success by creating new insights, and then show the value of creating organizational knowledge.

Reviewing a Typical Project

According to the Project Management Institute (PMI) there are little over 16 million project managers in the world.[1] By comparison, the research firm International Data Corporation (IDC) estimates that there are a little over 18 million software developers.[2] That means there's almost one project manager for every software developer. For many developers, that doesn't come as much of a surprise.

In the software development world, project management and software development seem to go hand in hand. Most software developers speak fluent project management. They may ask questions about software requirements, or whether a new feature is outside the scope of the project. They may even work with Gantt charts or other project management plans. These developers have internalized project management. They think about requirements, scope, and schedule as part of software development. Even though development can happen in many different ways, it's a natural tendency for these developers to follow project management practices.

That can be a pretty big challenge for your data science team. If members of your team have worked on traditional software projects, they might try to apply those project management principles. Data science teams don't work on projects. Remember that these teams are exploratory. That's the science in data science.

---

[1]Project Management Institute. *The PMI Project Management Fact Book; Second Edition.* Project Management Institute, 2001.
[2]"2014 Worldwide Software Developer and ICT-Skilled Worker Estimates," December 2013, accessed August 5, 2016, http://www.idc.com/research/viewtoc.jsp?containerId=244709.

Project management is a defined process. It requires you to have an understanding of the deliverable before you begin. Typical projects demand that specifications are established up front. They focus on delivering within a scope, schedule, and budget. You can't effectively manage these projects without some sense of these constraints.

A typical project also delivers a product or service. There should be a noun at the end of your project. Maybe you're completing a report, or your team delivers a software product. At the end, your project has to deliver something so that you know it's complete.

Let's look at a typical project. Let's say that you need to purchase a new server for your running shoe web site. You have a project manager working on the project. The first thing the project manager does is create a project charter, which is a quick one-page document that says what the project will accomplish.

If the charter is approved, the project manager creates a plan. The plan documents the project's scope, cost, and schedule. In this case, the scope of the project is purchasing a new server. The project manager estimates the cost, and assigns a date for the server to arrive. The project ends when the server arrives.

The scope, cost, and schedule are all balanced constraints. If the plan changes, the cost will most likely go up and the schedule will probably be extended. If the project manager decides to overnight the server, the cost will go up and the schedule will shorten. The scope, cost, and schedule are all balanced in this iron triangle.

So what does this have to do with data science teams? The short answer is nothing. Good project management is an entirely different discipline. It has different goals and processes. That doesn't mean that project management will be absent from data science; there's a good chance that someone will drive your data science team to use project management principles.

There's an old joke that if you have a shiny new hammer, everything looks like a nail. The same is true with most organizations and project management. It's the tool they're used to using. They don't feel comfortable spending money without their hammer and nails.

I once worked for an organization that was deeply invested in project management. The project management office was one of the most powerful departments in the company. The data science team struggled with this in nearly all their meetings. Stakeholders would ask about the scope of the project, and the data science team never had a good answer because they were trying to create new knowledge and find insights. The stakeholders would also ask when the team was scheduled to deliver those insights. Again, the team didn't really have an answer. They didn't know what they would find. They were still looking through the data. They couldn't set a date on when they would find their most valuable insights. That never left the stakeholder very satisfied.

If you're working on a data science team, you will almost certainly run into questions like these. If it's just a few project managers asking these questions, it probably won't be too much of a challenge. If the person who's sponsoring your project is asking these questions, you might have a real issue.

The best thing you can do under the circumstances is to communicate the difference between project management and data science. At the very least, be sure that everyone on your data science team understands the distinction. Try to stay away from project management language like scope, cost, and schedule. Over time, your audience might accept the different approach.

Working on a Data Science "Project"

Project management has been very successful in most organizations. It's been a shiny hammer that's helped nail down costs and manage schedules. It's been successful enough that organizations use project management for many of their efforts. That doesn't mean that project management is appropriate for all these efforts.

Data science is remarkably different from project management. Your team might explore new opportunities. They might be trying to make your data more accessible to the rest of the organization. Maybe they're looking for ways to better understand your customer or trying to detect security breaches or fraud. The team might even explore massive data sets from sensors or machines. These efforts don't fit into a typical project management framework. For one thing, you won't have a scope. Data science projects are exploratory. That's the science in data science.

You can't create a detailed description of what you're going to find before you start exploring. The whole purpose of exploring is that you don't know what you're going to find. You have to be able to react to your data. You need to expect the unexpected if you want to explore and gain new knowledge. In general, data science looks for new opportunities or tries to address current assumptions. It focuses on knowledge exploration and tries to deliver insights.

Think about the things that you do in your life that are more exploratory and empirical. Have you ever walked along a strip of restaurants and looked at menus? That's an empirical process. You're exploring each of the restaurants and responding to the menus. That's much different than if you made reservations at a well-known restaurant. Then you'd be planning where and when you're going and probably what you're eating.

Now, imagine that while you are exploring, someone asks you to commit to what you'll eat, what you'll spend, and when you'll be finished. Chances are, you wouldn't be able to answer. If that person insisted that you answer, you would probably jump to the first restaurant, look at the menu, and make an estimate. In a very real sense, you have to stop exploring. Instead of learning, you'd be planning.

That's exactly what happens when project management practices are applied to data science. Table 11-1 compares a typical software project to a typical data science project.

Table 11-1. Software project and data science project comparison

Typical Software Project	Typical Data Science Project
Develop a new customer self-help portal	Better understand customers' needs and behaviors
Create new software based on customer feedback	Create a model to predict churn
Install a new server farm to increase scalability	Find new markets and opportunities
Convert legacy code into updated software	Verify assumptions about customer use

I once worked for an organization that insisted on applying good project management practices to all their work. The data science team was no exception. The team tried to accommodate this practice by creating knowledge milestones and insight deliverables. In the end, it was completely unworkable. The knowledge milestones were just best guesses and discouraged the team from looking at anything interesting. They only looked for things that were easily proved or bordering on obvious because of the time constraints. Anytime I tried to ask more interesting questions, they were afraid they would miss a milestone.

As mentioned earlier, project management practices are beneficial to most organizations. Unfortunately, for your data science team, those practices have a chilling effect. Project management discourages uncertainty. It forces the data science team to only try and verify what is already known. If they find anything unexpected, it's seen as a bug and not a feature.

When you create milestones and deliverables, you're telling the team that they have a set time to verify what's already known. They measure their success by not finding new things. That's the opposite of what you want your data science team to do. You don't want to think of data science as a project delivering a product.

Comparing Project Challenges

Traditional projects rely on set requirements and careful planning. Remember that a typical project has scope, cost, and schedule. This isn't really compatible with the scientific method used with data science teams. There's no concrete deliverable to manage and you can't really balance these constraints.

Instead, data science teams are empirical and exploratory. These projects involve learning by looking. If you insist on a project plan, you're boxing the team into looking for what they already know. It's tricky to imagine most teams finding a lot of new data in a well-defined box.

If you think about the meetings in most organizations, they usually revolve around planning and hitting objectives. The language of most organizations still hinges on phrases such as mission, objectives, and outcomes. It's difficult to step back and imagine a team of pure exploration. For most organizations, it'll be a difficult transition.

So let's look at a project and compare it to a data science team. Then let's see what would happen if you applied planning and objectives.

Let's start with a typical software project. Your organization wants to develop a new customer self-help portal. The project charter is to create the portal as a way to lower costs. The project will have a set cost, but the organization will save money in customer service. The project has a strong return on investment (ROI). The plan lays out all the features in a requirements document. There's an estimate of the development schedule and all the costs are documented. All of these are outlined in the project plan. The project manager will update the plan throughout the project and help balance any changes.

Now let's imagine the data science team. It's a small team of four people. There's a research lead, two data analysts, and a project manager. They're tasked with better understanding customers' needs and behaviors. The leaders of the organization feel that if they can better understand their customers, they can convert that understanding into future revenue.

The research lead starts by asking a few questions:

- What do we know about our customer?

- What do we assume about our customer?

- Why does our customer shop with us instead of our competitors?

- What might make our customers shop with us even more?

The research lead would work with the data analyst to break these down into reports. Maybe they could create reports on the customers' income, as shown in Figure 11-1. They could also analyze social media platforms and create a word cloud of feedback from thousands of customers, as shown in Figure 11-2. For example, some of the largest words in the word cloud were "travel," "recipe," and "restaurant." The team could go back and ask more questions. Why do our customers like to travel? Where are they going?

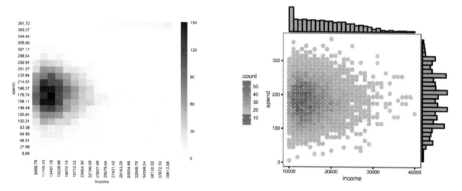

Figure 11-1. Income and spend

▨ **Note** The x-axis is the income and the y-axis is the spend. People with higher incomes don't necessarily spend more. Those who have an income around $20,000-30,000 seem to have the highest spend. See how to create this chart at http://ds.tips/n6cEc.

Figure 11-2. Word cloud

▓ **Note** See how to create this chart at `http://ds.tips/k8wRa`.

You can see how knowing more about your customer can lead to higher sales. Maybe you can work with the marketing team to advertise in travel magazines. Maybe you can start selling products that are closely related to traveling.

On the other hand, you might also discover that the whole exploration was a dead end. Maybe your data analyst created a report of where the customers were traveling, as shown in Figure 11-3. It turns out that a lot of your customers do travel internationally, but not enough to justify selling new products. So the team decides to drop it and explore other areas. Maybe next you'll try to explore your customers' favorite restaurants.

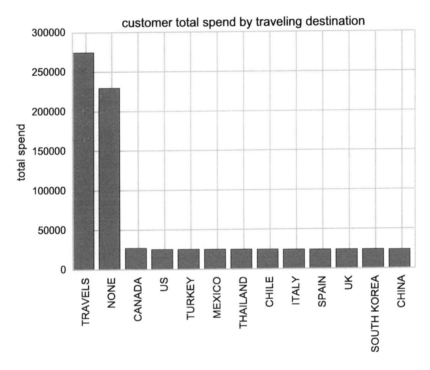

Figure 11-3. Where customers travel

▓ **Note** The total spend by customers who travel is greater than those who don't. However, if you compare the travel destinations, the total spend by each destination is less than those who don't travel. See how to create this chart at `http://ds.tips/y8seS`.

How would this fit into a traditional project management framework? What's the scope of your project? It's finding out about your customer. It's about new knowledge. How do you know when you're done knowing something? When will this new knowledge be delivered? What's inside the scope of the project?

All these questions would probably give any project manager a blank stare. Learning is a verb and not a noun. If you don't know what you're looking for, there's no way to measure what you'll learn. If you know what you're looking for, you're not really learning something new. What about schedule? Will they continue working on this project even after they've reached a dead-end? When does the team stop working? Finally, what about the cost? If you don't know how long the work team is working, how do you budget for their time?

As you probably noticed, this data science project will not fit into a project management framework. You'll see the same will hold true for most, if not all, of your data science explorations.

Defining Success Criteria

In a 1921 interview, Thomas Edison said that his assistant was discouraged by their failed experiments.[3] The famous inventor was cheery and assured him that they hadn't failed. They were always learning something new. Once they learned something, they would try a different way. Today, we know Thomas Edison was right because of his many successes. A few of them are still in use today. There were also a few experiments that were lost to history—not many play on their concrete piano. We have the benefit of looking at his legacy as a string of successful experiments. We don't see the failed experiments that took up most of his time. There were many more failures than successes.

If Edison were using modern project management, he would've run into a couple of challenges. How would he define his success criteria? You couldn't just look for things that worked. If you did, you'd need a lot of patience. His experiments would last for months or years before producing anything that even looked like a success.

We should look at data science success the same way as Edison looked at his experiments. Just ask the simple question: did we learn anything new? Your team will run many experiments on the data, and most of the experiments will be failures or dead ends. Try not to think of these as failures. Not every

---

[3]"Why Do So Many Men Never Amount to Anything?" by B. C. Forbes, [Interview with Thomas Edison], *American Magazine*, Volume 91, January 1921. Crowell Publishing Company, Springfield, Ohio. (Google Books full view Start Page 10, Quote Page 89, Column 2) http://books.google.com/books?id=CspZAAAAYAAJ&q=%22I+cheeril y%22#v=snippet&

experiment will lead to insights. It also may be true that most of your insights won't have very much value. Maybe you'll find out that most of your customers are pet owners. That might be interesting, but it probably won't have much value.

This approach may be challenging in many organizations. In some organizations, it might even be annoying. You'll know you're running into this problem when people ask, "What is that team doing?" or even worse, "What does that team do again?" This can be an organizational challenge. Managers may hire a data science team as an experiment to see what they come up with. This can make things more difficult for your data science team.

There are a few things that you can do to present some success criteria:

- Make sure that your team is as transparent as possible. Fight the urge to stay cloistered away from the rest of the organization. Often, if people don't understand what you're doing, it doesn't take long for them to question why you're there.

- Make sure you're trying to solve large problems. You want to keep your team ambitious enough to tackle interesting questions. If the questions are too timid, it might be difficult to show interesting results.

- Try to show what the team is learning through regularly scheduled storytelling meetings. In these meetings, cover the questions that the team is working on and provide a few recent insights.

I once worked for a university that hired a group of "unstructured data specialists." The provost wanted to have a data science team that looked for new insights. The team worked in an office near the administrators who hired them. No one else in the university knew what they were doing. Most didn't even realize they were there. The problem was that this data science team had trouble asking any interesting questions. No one in the rest of the university would take the time to meet with the research lead.

Things would've gone much smoother for this team if the team had been placed in a location close to the rest of the faculty instead of the administrators. That would have enabled them to work with everyone from the beginning to come up with interesting questions. They could've had storytelling meetings that gave insight into these questions.

If you're the research lead for a data science team, work hard to keep the questions closely connected to the rest of the organization. Stay transparent about what you find. Give frequent demonstrations of interesting insights. Try to draw on the rest of the organization so they understand the value in data science.

If you're the project manager for a data science team, try to work hard to make sure that the team is sitting with everyone else. Some of your best inspirations might be people dropping in and asking questions. The better connected the team is with the rest the organization, the easier it will be to create interesting questions.

Summary

In this chapter, you reviewed a typical project in an organization. Next, you found out what it's like to work on a data science project. Then you learned about the different project challenges for each type of project. Finally, you found out how to define the success criteria for your data science team and what you need to communicate to the rest of the organization. In Chapter 12, you'll find out how to use a data science life cycle.

Using a Data Science Life Cycle

Most of the people on your data science team will be familiar with a typical project life cycle. People from a software development background are familiar with the software development life cycle (SDLC). People from data mining probably used the Cross Industry Standard Process for Data Mining (CRISP-DM).

Each of these life cycles works well depending on the project. The problem with these life cycles is that they require you to know a lot about what you're doing before you start. In software development, you have to have a clear scope. With data mining, you have to know a lot about the data and business needs.

Data science is empirical. You don't know what you'll find. You might not even know what you're looking for. Instead, you have to focus on interesting questions, and then create a feedback loop to make sure those questions are tied to business value.

© Doug Rose 2016
D. Rose, *Data Science*, DOI 10.1007/978-1-4842-2253-9_12

Nevertheless, a life cycle can be very useful. It's like a high-level map that helps keep the team on track. That's why for a data science team, you'll want to try a different approach. You can use a data science life cycle (DSLC) as a way to set some direction for the team.

In this chapter, you'll explore the SDLC and CRISP-DM so you can understand how they differ from a DSLC. Then you'll learn how to use the DSLC and how to effectively loop through the DSLC questions.

Exploring SDLC or CRISP-DM

You've seen how difficult it is to get data science teams to work within a project management framework, so let's look at two of the life cycles commonly used in project management. A life cycle is a series of steps you take when you're developing software or approaching a problem.

There are two life cycles that you're likely to run into in large organizations.

The first is the software development life cycle (SDLC). This life cycle has six phases, as shown in Figure 12-1. Underneath each phase is an example of what happens during that phase.

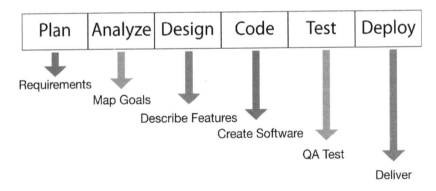

Figure 12-1. The Software Development Life Cycle (SDLC)

This is typically called the waterfall model because each one of these phases has to be complete before the next can begin:

1. **Plan and Analyze:** Plan the software and analyze the requirements.

2. **Design and Code:** Create the basic designs and start your coding.

3. **Test:** After the code is complete, the quality assurance people test the software.

4. **Deploy:** After it passes all the tests, it is deployed for people to use.

The second life cycle you're likely to see is the Cross Industry Standard Process for Data Mining (CRISP-DM), which is used for data instead of software. It's modeled to be a little bit more flexible than the rigid waterfall model. It also has six phases, as shown in Figure 12-2:

1. Business Understanding

2. Data Understanding

3. Data Preparation

4. Modeling

5. Evaluation

6. Deployment

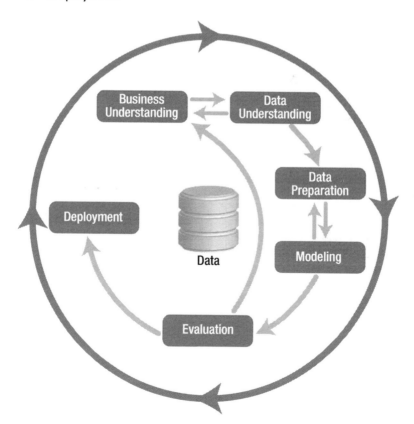

Figure 12-2. Cross Industry Standard Process for Data Mining (CRISP-DM)

What both of these life cycles have in common is that they're designed for "big bang" delivery. You spend a big chunk of your time either in the plan-and-analyze phase for software or the business understanding phase for data. The goal is to gather up as much information as you can before you start working. Then you deliver it all in the end with a big bang.

That's not necessarily the best approach when you're working in data science because of its experimental and exploratory nature. Imagine a typical data science project. Let's say that your data science team is identifying typical customer behavior before they decide to leave you for a competitor. Sometimes this is called the customer's churn rate. Your data science team might be able to clearly state their intent: to see what customers do before they leave and to create a model to predict when someone might leave.

However, your team will not be able to plan out their work. They may find their best models by looking through social network data, from the company's sales data, or even from a promotion that your competitor offers that becomes unusually successful. The point is that they won't know until they start looking.

Your team will spend too much time planning if they're forced to use the SDLC or the CRISP-DM process. They won't be able to apply what they're learning from the data. That's because they're forced to plan out their work before they even begin modeling or coding. Defined processes like SDLC or CRISP-DM require that every piece of work be understood. If you make a mistake, you have to deal with a change request in SDLC and a reevaluation in CRISP-DM.

If you want your data science team to be flexible and exploratory, you can't apply a standard life cycle. Instead, you should look for a more lightweight approach to delivering insights so you can have structure while still being flexible enough to adapt to new ideas.

Using a DSLC

Data science doesn't work very well with the existing process life cycles. It's not enough like software to fit the SDLC, and the data mining process of CRISP-DM is a little too rigid. That doesn't mean that a data science team should just work in whatever way feels right. There is real value in these life cycles. One value is that it gives you a high-level map of where you're going. This is really useful when you're starting a data science team. You get a general sense of the path forward, so you can start with the end in mind.

The danger with the life cycle is that it becomes the primary focus of the work. You want to use the life cycle as a vehicle for better data science. You don't want to follow the process for the sake of following the process. A good life cycle should be like a handrail. You want to think it's there for when you're going up and down the stairs. You don't want to cling to it with every step. After a while, you shouldn't even notice that it's there.

For data science projects, you can use DSLC. This process framework is lightweight and less rigid. DSLC has six steps, as shown in Figure 12-3 and discussed in more detail in the following sections.

This life cycle is loosely based on the scientific method.

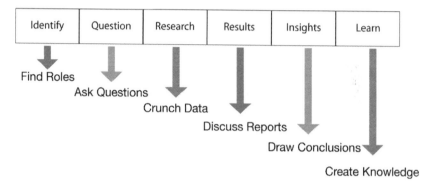

Figure 12-3. Data science life cycle (DSLC)

Identify

As a data science team, start by **identifying** the key roles in your story. In the end, you want to be able to tell an interesting story with your data. The best way to start a story is to identify key players. Think about it as a scene in a play. Who walks into the room? Is there a main character or protagonist? Is there a backstory that helps make sense of his or her actions?

Let's go back to the running shoe web site. Who are your key players? There's the runner. Maybe the runner has a partner who influences his or her running habits. Maybe your runner's partner is a doctor, a blogger, or a trainer. Each of these players could be a part of your data science story.

Question

After you've identified your key players, you can ask some interesting questions. Your team's research lead might start out by asking, "Is there a blogger who influences your runner?" Maybe the trainer plays a big role in influencing what your runner purchases. They might ask, "Are CrossFit trainers recommending our products?" These questions are a key first step in exploring your data. Remember that data science is experimental and exploratory. When you start with a good question, you're more likely to have interesting research.

Research

The data analyst wants to work closely with the team to try and get at some strategies for researching the questions. The team decides to explore the relationship between the runners and their partners. Here, the research lead would ask the data analyst how they could get this information.

How can you determine if someone's a running partner by the data on the web site? Maybe you could send find-a-friend promotions to the same ZIP code. The data analyst could try to cross-reference the customer data with people who are friends on social network sites. If the data analyst can't research this question, the team could come up with strategies for the future. Maybe the web site should create a special promotion for running partners.

Results

After you have your research topic, you want to create your first reports. These results are for the team. They should be quick and dirty. Hopefully, your data science team will go through a lot of questions and a lot of reports. Most of these will be duds. They might be interesting, but not interesting enough to explore further. You don't want your data analyst spending too much time perfecting the results.

Insight

Finally, your data science team should look at the results to see if there are any interesting insights, as shown in Figure 12-4. Maybe the data suggests that most of your customers run with partners. That insight might be very valuable to the marketing team.

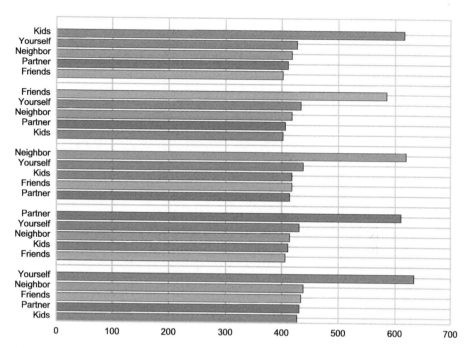

Figure 12-4. Who people run with

▒ **Note** This chart was derived from quarterly surveys where customers selected who they run with. The first bar in each grouping is the tally for the people they run with the most. Below that are tallies for other types of people survey respondents run with. For example, people who most often run on their own also run with their neighbors, and people who run with their kids will otherwise prefer to run on their own. See how to create this chart at http://ds.tips/cH6th.

Learn

In the end, your team will bundle up a bunch of these insights and try to create organizational knowledge. It's here that your team will tell the customer's story. You might want to use data visualizations to back up your story. This new knowledge is what really adds value to the rest of the organization. If you tell a compelling story, it may change the way your organization views their business.

Looping Through Questions

The software and data mining life cycles don't have enough flexibility to deal with new information. That's why you use a DSLC, which is much more light-weight. If your data science team finds something new, they shouldn't have to fight the process to have your organization benefit.

As mentioned, the DSLC has six areas: Identify, Question, Research, Results, Insights, and Learn. To summarize, first, you identify the players, and then you create some interesting questions. Your data science team then should agree on how to research those questions. You'll discuss the results and see if there are any insights. Then you collect your insights and create a story and tell the rest of the organization what you've learned.

These six areas are not like the software development life cycle, where each step leads to the next. Instead, think of the three areas in the middle as a cycle. Your data science team should cycle through the questions, research, and results (see Figure 12-5).

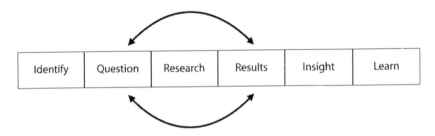

Figure 12-5. Cycle through question, research, and results

This cycle of questions, research, and results is the engine that drives your data science team. Each of the three roles on your team focuses on one of these areas. The research lead focuses on creating the right questions. The data analyst works with the research lead to come up with the right research question and create reports. Then the project manager communicates those results to the rest of the organization.

Let's go back to the running shoe web site example. Every few years, there's a running shoe that's a breakout hit. Your manufacturers produce hundreds of shoes, but every so often, one does much better than all the rest. You approach the data science team and ask them to create a predictive model for these top shoes, as shown in Figure 12-6. The research lead asks an interesting question: "what about a shoe makes it a hit?" She works with the data analysts to come up with some interesting questions:

- Is it the color of the shoe?

- Is it some new technology?

- Was the shoe featured in a magazine and benefited from a network effect?

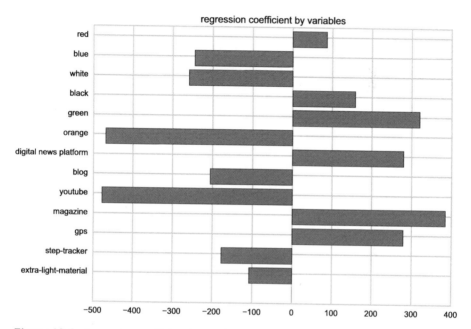

Figure 12-6. Regression coefficient by variables

▒ **Note** Using sales value as y variable and colors, marketing channels, and technologies as x dummy variables, we can see the top-selling products are most likely to have black but not white, and orange also helps. You can interpret the data to say: If everything else stays the same, if the product is advertised on digital news platforms, it's likely to help its sales by $200 versus advertisement on a blog, which will decrese sales by close to $200. See how to create this chart at http://ds.tips/wrA3e.

The research lead, data analyst, and project manager cycle through these questions, research, and results. Most of the questions and results may be duds. A few of them may lead to insights. Maybe the results suggest that it's a combination of some key attributes to use as insights. The team then bundles this up and tells a story, which is new knowledge. The story is that the best way to predict a hit is a combination of media buzz, new technology, and interesting design.

Now, the running shoe company can use this knowledge to create a new product. Instead of creating hundreds of shoes waiting for a hit, the company can supply the specifications and environment to create a hit.

Some organizations that have strong data science teams are already using this approach. The subscription service Netflix used this approach to create their hit series *House of Cards*. They had 33 million subscribers at the time. Their data science team looked at what customers were watching, ratings of shows, what plot viewers liked, and popular actors (Kevin Spacey was very popular). Netflix determined that political shows were very popular and hired Spacey. Then they modeled the new show on the popular British version of the program. They created a predictive model based on what made a popular show. They worked to cycle through questions, research, and results. The data science team then created a story of what their customers would like to see. That story became a plot that turned into a hit television program.[1]

This cycle of question, research, and results drives your insights and knowledge. Your data science team loops through these areas as part of the larger DSLC. Remember to not think of this life cycle as a waterfall-style process. Instead, think of it as a few steps to start and then a cycle in the middle to churn out great stories.

Summary

In this chapter, you saw the SDLC and CRISP-DM so you can understand how they differ from a DSLC. You learned that both SDLC and CRISP-DM have six phases and you explored those phases. You found out how to use the DSLC to effectively loop through valuable insights. In Chapter 13, you'll learn how to work in sprints. With sprints, you'll be able to frequently show your stakeholders something valuable and build up organizational knowledge.

---

[1]David Carr, "Giving Viewers What They Want," *The New York Times*, October 15, 2014, http://www.nytimes.com/2013/02/25/business/media/for-house-of-cards-using-big-data-to-guarantee-its-popularity.html?_r=0.

Working in Sprints

One of the key ways to stay adaptive is to break down your work into smaller chunks, so you're frequently showing your stakeholders something interesting. Your team can create questions and get quick feedback to see if the data stories are valuable and resonate with the rest of the organization. If it's not valuable, the team can quickly pivot to other questions. If it is, the team can take a deeper dive and maybe even come up with questions based on feedback from the business. This feedback loop is essential in making sure that the team's work is tied to business value.

In this chapter, you find out how to work your way through DSLC sprints, create a question board, focus on a few meetings, break down the work, and tell an interesting story. All of these skills will help you work through the DSLC more effectively.

Iterating Through DSLC Sprints

It's important to remember that the entire DSLC should be run in quick increments. The DSLC isn't designed to run over weeks or months—it's small enough to fit into two weeks of work. Every two weeks, the team can refine their work, create new insights, and come up with improved questions. If the business decides that the work isn't valuable, the team can change course and try something new.

© Doug Rose 2016
D. Rose, *Data Science*, DOI 10.1007/978-1-4842-2253-9_13

Now that you've seen the DSLC, you might be wondering how this looks in practice. One of the most important things to remember is that the DSLC is not structured to run in phases like the SDLC, where you don't usually start the next phase until the previous phase ends.

The DSLC is about making sure you concentrate on the six areas: Identify, Question, Research, Results, Insights, and Learn. The whole life cycle should run in a short "sprint." You may have heard of the term sprint, which is widely used in agile software, but it actually came from product development. A sprint is a consistent period of time where the team runs through an entire life cycle. Each sprint should run through all six areas of the DSLC.

The data science team should run in a two-week sprint. That's long enough to find insights, yet still short enough to adapt to new ideas, as illustrated in Figure 13-1.

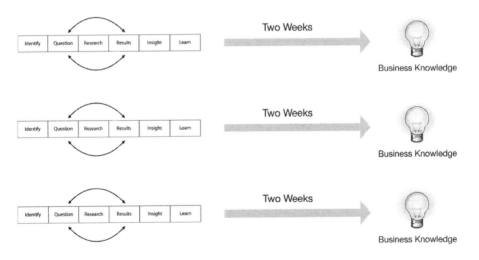

Figure 13-1. Two-week sprints

The main advantage of running in sprints is that it shortens the time between "concept to cash." Many organizations take a long time to come up with new ideas, and then go through a long delivery process. These new concepts might not add value for many months, and certainly won't add any new cash to the business until well after they come up with the original concept. A two-week sprint adds value more frequently. Even when there are no insights, there will still be finished questions.

You don't want your data science team working in a long timeframe. If you work in months or quarters, too much time will pass between questions and results. By the time you get any insights, the data may have changed.

You also don't want your team investing too much time in any one question. Remember that a vast majority of your research will lead to dead ends. Most of your questions will be duds. You have to open up a lot of oysters before you find any pearls. That's why you want to burn through these questions quickly, so when you find something interesting, you can build off of that work.

I once worked with a state's Department of Education that was trying to better understand the students who were attending their public schools. They had a large data science team. The team was trying to create a model to better predict the students' interests. Unfortunately, they tried to use the SDLC. The team tried to work in sprints, but preferred to work in a typical waterfall.

The data science team spent most of their time planning. When they started their project, they spent several weeks in brainstorming meetings. If you've ever been in a brainstorming meeting, you know it needs to be well structured. If it's not, the meeting goes well beyond its scheduled time. That's exactly what happened.

In fact, several months went by before the team was able to publish their first requirements document. The document listed several questions that they wanted answered and outlined the research techniques. Then the team presented the insights they hoped to gain from their research.

There were many challenges with this approach. Remember that data science is exploratory and experimental. They only laid out a few experiments, and then they recorded what they hoped to learn. They set out to prove what they already assumed to be true, and since the proof was a requirement, it meant that if they turned out to be wrong, the project failed.

It would've been much more effective if the data science team decided to run in sprints instead. This would've allowed them to come up with more questions and run more experiments. Instead of hoping to be right, they could've focused completely on learning something new. It also would've been far more efficient. After three months working on the project, the only thing the team had to show for it was a list of questions and a few assumptions. Within those three months, they could've gone through a dozen smaller sprints and known far more about the data because they would've gone through many more questions and experiments.

If you're on a data science team, try to remember to not get overwhelmed by the data. Don't create big plans over several months. Instead, try to keep things small and simple. Build up your knowledge bit by bit, so you can adapt to new ideas as opposed to being focused on any one path.

Creating a Question Board

When you're working on your data science team, your research lead is responsible for driving interesting questions. Coming up with good questions is not an easy task. A good question can stir up a lot of new information and force people to rethink their work. That's why most organizations tend to shy away from good questions. When you have a good question, it can cause some irritation. You're almost itching to find the best answer. That can lead to a lot more work and sometimes even more questions.

Even today, most organizations still try to focus on improving what they know. They figure if they can optimize, they'll always be ahead of newer competitors. A good question can often shake up these well-ordered plans. Good questions have a tendency to break up predictability and can turn a well-ordered set of objectives into an open-ended question.

It's the research lead's responsibility to breakup this well-ordered process and inject some exploration and experimentation. One of the best tools a research lead can use is a question board. A **question board** is usually a large white board filled with sticky notes, typically positioned near the data science team. There should be lots of space with new questions and a short stack of sticky notes in one of the corners. You might want to create a large arrow pointing down to the stack of sticky notes. Some teams will add the caption, "Ask a question."

The question board is used to solicit questions. The research lead drives the questions, which doesn't mean that she comes up with all of them. It should be a combination of her own ideas, the data science team's questions, and open-ended questions from the rest of the organization.

The question board should be open and inviting. Try to make it look as enticing as possible. You want anyone to be able to walk by, grab a sticky note, and create a quick question. Make an effort to keep it lighthearted and fun. Some teams even make it almost like a game. They put a big bowl of candy next to the question board or they print out a sign that says, "Ask a question and win a prize."

Note See Chapter 16 for more about how to organize your question board and how to use different color sticky notes for different types of questions.

The question board also helps everyone in the organization understand the purpose of the data science team. When your data science team does the storytelling presentation, people will often recognize their own questions and will be more likely to ask questions in the future. They may even encourage their coworkers to ask questions as well.

You can never have too many questions. The research lead works with the rest of the team to prioritize the most interesting ideas. If you get your organization to use the board, it starts to look a little bit like a three-dimensional search space. You can see patterns in what people ask. The board itself becomes another data source.

I once worked for an organization that put up a question board in the corner next to the data science team. At first, it was just a curiosity. People would just come by and read it the same way people are drawn to an announcement board. The team was smart and put it next to a water cooler. After a while, a few new questions popped up on the board. They were mostly silly and didn't have much value. Still, the research lead used the question board for communicating what the data science team was doing. The team posted their questions and continued to give presentations.

Over the summer, this organization brought in a whole new group of student interns. For the first month, the students were trying to figure out the business. Being students, they were a little bit more comfortable asking questions. The board started to fill up with their sticky notes. Some of the questions they asked were very intuitive. They were looking at the business from a fresh perspective. The questions were so simple and well structured that the data science team started making them the highest priority. They helped the team explore the business in interesting new ways.

If you're the research lead, be sure to take advantage of the question board. It's a simple way to get interesting new questions while at the same time communicating your progress to the rest the organization.

Focusing on Just a Few Meetings

Your data science team will typically want to work in two-week sprints. The team will have a lot to do, so they need some structure to stay efficient. Remember that you'll go through every area of the DSLC during each sprint. To work at that pace, the team needs a set amount of time to work, and can't attend many open-ended meetings. They have to account for all their hours.

Each meeting will need a set time box. A time box is pretty much what it sounds like: a set amount of time that the team agrees upon before the meeting. Let's say your team has a meeting with a one-hour time box. Whatever they decide at the end of that time-box will have to last until the end of that sprint. You can never reschedule or follow up on time-boxed meetings. They start and then they end.

In most organizations, meetings aren't necessarily bad. They're a good way to bring up issues and reinforce culture. The challenge with meetings is that they add a lot of unpredictability to your week. Your data science team needs a predictable schedule so they can commit to a certain amount of exploration and discovery. You want your data science team working at a sustainable pace.

The data science team should have the following five meetings during each sprint:

- Research Planning
- Question Breakdown
- Visualization Design
- Storytelling Session
- Team Improvement

These five meanings work together to help deliver all the areas in the DSLC, as shown in Figure 13-2. Each one of these meetings is time boxed.

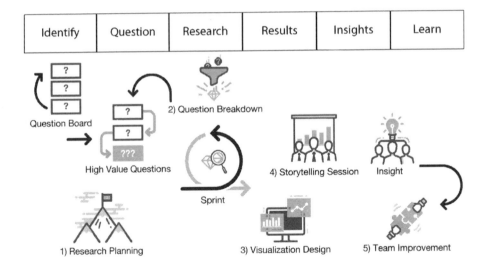

Figure 13-2. Five meetings for the DSLC

Research Planning

The team should start each sprint with their Research Planning meeting. Some teams choose to have their sprints start on Wednesday, which enables them to deliver a Storytelling Session on a Tuesday afternoon. It's much easier to get people to attend these meetings on a Tuesday than a Friday. The Research Planning meeting is when the team decides which questions they're most interested in exploring in the next sprint. It's typically time-boxed to two hours. In the meeting, the research lead and the data analyst work together to come up with that week's research agenda.

Often the analyst will have to wrangle a lot of data to even attempt to get at an interesting report. The research lead and data analyst will work together to create a minimum viable report. Maybe they don't need to scrub all the data to start experimenting and exploring. There should be a natural tension between the data analyst and the research lead. The research lead will want to create small, quick-and-dirty reports. The data analyst will want to scrub large data sets and tackle bigger problems. This meeting is designed to help team members come up with a compromise strategy. What's the least amount of work you need to do to prepare the data and create an interesting report?

Question Breakdown

During each sprint, the team will want to have at least two one-hour Question Breakdown meetings. In these meetings, the whole data science team will try to come up with interesting questions and put them on the question board. They'll also take any new sticky notes off the question board to see if they're interesting.

They'll also look for question themes:

- Are there any common questions?
- Are there large questions that can be broken down into smaller, more manageable ones?
- Did anybody respond to any of the team's questions?

The team will work together to try and prioritize some of the more interesting questions for the next sprint.

Visualization Design

The third meeting is the Visualization Design meeting. This meeting is usually time boxed to one hour. This is when the research lead and the data analyst work together to create an interesting visualization. It's usually just a rough draft of what the data analyst will use during the storytelling session.

Storytelling Session

The fourth meeting is the one-hour Storytelling Session. This is when the data science team presents a story about what they learned during that sprint. In this meeting, they show some of their nicer data visualizations, talk about questions on the question board, and then tell stories about those questions or ask their own.

Improvement

Finally, at the end of each sprint, the team should have a two-hour Improvement meeting to evaluate their progress and discuss if they're working well together and if they should make any changes.

All of these meetings should work together so the data science team can deliver interesting insights after each sprint. Remember that there's a lot to do during each sprint, so keep these meetings time boxed so you can focus on discovery.

Breaking Down Your Work

By now, you know the difference between the SDLC and the DSLC, and you know that the DSLC is best delivered in two-week sprints so you can break down the work and deliver valuable insights. When you're on a data science team, there are always large data sets that need scrubbing and new sources to explore. In fact, most of what you'll be doing is preparing your data. When you work in sprints, you're forcing the team to do the minimum amount of preparation.

Doing the minimum amount of data preparation might sound like a bad thing. Most people want to do higher quality work. In actuality, when you do the minimum amount of preparation, you force your data science team to focus on insights and not just capability. You don't want your team spending weeks or even months just setting up the data. Instead, you want the team to almost immediately start exploring the data.

You also have to look at it from the organization's perspective. Most organizations aren't really interested in the data. They are interested in the knowledge they gain from the reports. From an organization's perspective, managing that data is part of the cost and not the benefit. This means that there can be increased pressure to extract value from the data as quickly as possible. It's difficult for data science teams to spend too much time prepping data and only deliver reports at the end.

In many ways, this is similar to how many organizations now view software. In the beginning, most organizations viewed software development as a bit of a mystery. They left most of the details to highly skilled software engineers. These engineers would spend most of their time planning and preparing for a big release. Now most software developers are forced to deliver valuable software in much smaller chunks. They spend less time preparing and more time delivering. This allows the organization to get a look at the value before the team gets too far along.

Like early software development, in many organizations, data science is still a little bit of a mystery. The team still gets a lot of leeway in how they want to do their work. However, it won't take long for managers to start asking tougher questions. Currently, most data science teams have the luxury of spending a lot of time preparing large data sets. Once those managers start asking questions, the team will instead have to focus on the minimum viable data prep.

I once worked for an organization that was focused on automating the process of scrubbing a very large data set. They wanted to plug it into an even larger data set that they already had housed on their cluster. For months, the science team was solely focused on this task. They downloaded open-source software tools and purchased some commercial products to help them prepare the data. After several months, they had created several scripts, software tools, and practices that automated the process of moving these large data sets into their cluster.

After they moved it over, they had a meeting with the vice president of enterprise data services. They showed a PowerPoint presentation of how much data the cluster had consumed. They went through several slides of how difficult it was to scrub and import the new data set. Near the end of the meeting, the vice president asked an interesting question. He simply asked, "What do we know now that we didn't know before?" The question landed in the room with a thud. It was clear from the silence that no one had thought about the data that way for months. Everyone in the room was completely focused on capability. They had forgotten the real value for the organization.

If they had delivered in two-week sprints instead, they could've focused on the value much more quickly. Instead of building out the entire data set, they could have worked with smaller subsets of the data to immediately start creating reports and exploring the data. When you explore the data, you get a much better sense of the value. You're in danger of having the work become a routine when you just focus on scrubbing and importing the data. It's as if you spent all your time setting up the table for your dinner party and didn't leave any time to prepare a great meal.

Telling an Interesting Story

There's a big difference between presenting data and telling a story. For one, telling a story is much more challenging because you're doing a lot more work by bringing in the data and what you know about the business and throwing in what you know about the world.

When you put up a PowerPoint presentation with a data chart, you're saying, "Here's what I see." When you tell a story, you're saying, "Here's what I believe." This is a lot more difficult, and in a way, it's also a lot more personal. That's what makes storytelling such a valuable skill.

▓ **Note**   See Chapter 19 for more about how to tell an interesting data story.

When you tell a story, you do several things at once. First, you simplify the complexity in your data. You also define the roles of the people who were involved in creating this data.

Next, you bring in your own knowledge about the organization. This could be through your experience or research. You take a simple observation about people and data and then put it in the context of the organization. You don't just use the data to talk about where and what—you also present the why.

The third thing you do is make your data more memorable. Most studies show that when you present something in a PowerPoint format, very little of the information gets through to your audience. Those bullet points might be easy to create, but they're just as easy to forget. A story more effectively captures your audience's attention. If you can weave a good story, you're more likely to get everyone engaged.

Finally, a good story will have a call to action. It will either tell you something new or justify your continuing to look. If you can tell a good story, include your audience in the exploration. You'll be much less likely to have someone ask, "Why are you guys doing this again?"

Let's go back to our running shoe web site. Imagine that your data science team has been working on the question about increasing sales. You work together with the team to break the question down into several smaller questions. One of these smaller questions is, "Are people buying things on their wish list?"

The research lead and the data analysts work together to create a quick and dirty report to see how many wish list items were converted to purchases. Then they create a time series to see if these purchases were going up or down. Typically, the team would get together for a Visualization Design meeting the day before the Storytelling Session. In this meeting, they would try to convert the raw data and ugly reports into a nice visualization, and then use that visualization to tell an interesting story.

The data shows that in the summer months, people are more likely to convert their wish list items to purchases. That's just the raw data, but it's not a very interesting story. Why are people interested in shoes in the winter, but waiting until the summer to buy them? The data science team decides to tell a story.

They use the title, "Summer Dreamers: Why Do Winter Shoppers Buy Shoes in the Summer?" Next, the data analyst uses the whiteboard to come up with a first draft of the data visualization (Figure 13-3).

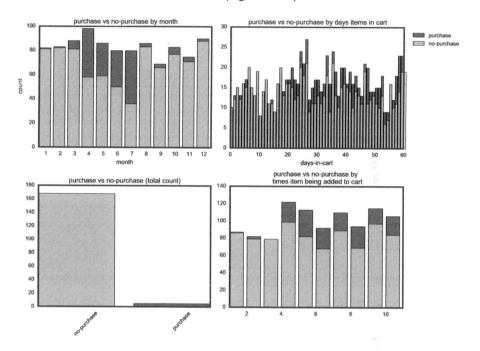

Figure 13-3. Purchasing habits throughout the year

On the upper left, there are more people purchasing items in their shopping cart during the summer months (June-September), but on the bottom left most items in the shopping cart did not get purchased. On the top right, items that are bought are left in the shopping cart for more than 20 days and being repeated added back more than four times. See how to create this chart at http://ds.tips/truD6

Notice how the story already makes the data more interesting. Imagine if the data analyst used the title, "Annual Wish List Conversion Rates" and included the simple time series graph on a PowerPoint slide. Something like that wouldn't pique anybody's interest. There's no context or a call to action.

The next day, the data science team uses their "summer dreamers" visualization to tell an interesting story to the rest of the organization. The story says that many of your customers are thinking about running in the winter, but they're only buying shoes in the summer. This story encourages further questions. Are people running in old shoes during the winter? Are they just not running? Do they not need new shoes because they're mostly running indoors? Should we make a special running shoe designed for indoor running?

Hopefully, you'll get these types of questions during the Storytelling Session, and then you can add them to your question board. If you can tell a good story, everyone in your organization will want to take part in your discovery.

Summary

In this chapter, you saw how to work your way through DSLC sprints. The DSLC has five core meetings. In these meetings, your team will work with a question board, break down the work, and tell an interesting data story. In Chapter 14, you'll learn how to avoid pitfalls you may encounter while working in data science sprints.

Avoiding Pitfalls in Delivering in Data Science Sprints

In this avoiding pitfalls chapter, you'll find some ways to change the mindset of your company. You can start by imagining how to work without objectives. Many organizations focus on objectives and create powerful compliance departments. These departments ensure that everyone meets those objectives. This focus can keep your team from exploring and discovering. A data science team needs to take advantage of serendipity and add to organizational knowledge. They need to emphasize exploring over routine work.

Changing the Mindset

Many companies underestimate the organizational change when starting their data science teams. It's not as simple as having a group of statisticians looking at the data. It's about treating your organization's data in a different way. It's

© Doug Rose 2016
D. Rose, *Data Science*, DOI 10.1007/978-1-4842-2253-9_14

no longer a static resource that needs to be controlled. Instead, it's an ever-changing source of insights that can't be disregarded.

▓ **Note**   See Chapter 25 for more about starting an organizational change.

Part of changing this mindset is letting go of strategies that may have worked well in the past. If you want to explore, you have to get rid of project objectives and planned outcomes. These are often barriers to discovery. You have to be comfortable with the idea that you don't know where the data may lead. You may even rely on simple serendipity.

At the same time, you can't wander without any purpose or direction. You need to create organizational knowledge that adds real value. You have to be open and exploratory, while still being practical and driven by business value. If you're too open, your team might get lost in the data. You'll explore dead ends and ask questions that no one is interested in answering. This will frustrate your stakeholders and limit your success. On the other hand, if you're too focused on objectives, you won't discover many new insights. You'll be limited to what you already hope to be true. There won't be much room for new discoveries.

If you want to be both explanatory and driven by business value, you need to change how you think about your work. The first step is taking a hard look at an old friend: planned objectives.

Working Without Objectives

Your data science team will want to use the DSLC and sprints so they can tell interesting stories every couple weeks. These practices help your team explore the data and ask great questions, and should help get your team focused on exploration. Still, for many teams, the biggest challenge is trying to change the organization's mindset.

As mentioned at the beginning of this chapter, most organizations still view work as a series of goals and objectives. That's why most key roles focus on management and compliance. A project manager ensures the team complies with a project plan. A lead developer helps maintain certain coding standards. A quality assurance manager enforces standards like Six Sigma. Even the CEO is focused on setting clear goals for the rest of the organization to follow. All of these popular roles center on compliance. They make sure the team stays true to their objectives. The people in these roles tend to be very influential. Chances are, they will try to apply this objective-driven mindset to your data science team.

That's a real challenge for the data science team, because they are about exploration and use an empirical process to research and learn from the data. It's difficult to work off of typical objectives. Exploration by definition is about searching and finding something unfamiliar. Objectives are about staying true to your intended purpose.

You can certainly mix exploration and objectives. If you find yourself in a new city, you might have an objective to find something great to eat. You then explore the area looking for good food stands or restaurants. You have the objective of finding dinner, but you're still open to exploring new ideas. The problem is that most organizations aren't quite this flexible. They tend to narrowly define their objectives. The objective itself becomes the top focus. A team isn't celebrated for changing course after they find something new. A successful team typically has a well-defined objective and meets their goals within the expected time frame. This focus on objectives can create a very difficult environment for exploration.

Let's go back to our running shoe web site. Imagine that your data science team is given the objective of creating a report that breaks down purchases by various credit cards. You want to see if accepting different credit cards might increase sales. While the team is exploring the data, shown in Figure 14-1, they notice something unexpected. It looks like there's a positive correlation between shoe sales and customer ratings. You might expect that the shoes with the highest ratings might have higher sales. The data science team notices, however, that shoes with any ratings had higher sales. The lowest-selling shoes were the ones with no rating at all.

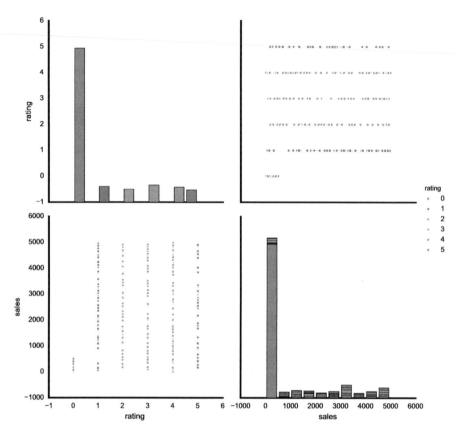

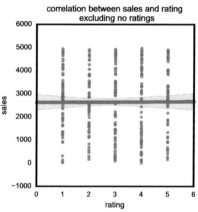

Figure 14-1. Correlation between shoe sales and customer ratings

---

▨ **Note**   Assume that 0 is no rating and the rest are the real ratings. Most items with total sales dollars close to 0 have no ratings. The rest of the data, however, does not indicate that higher ratings yield higher sales. In fact, the reason might be the opposite. Because the sales are higher, more people who actually bought the item are willing to leave feedback on the web site. In the bottom chart, however, if we ignore those that have no ratings, there's almost no correlation at all. See how to create these charts at http://ds.tips/8refR.

Based on this data, the team pivots to take advantage of this new discovery. They create an entirely different set of reports that correlate ratings with top shoe purchases. At the next Storytelling Session, they talk about how the customer is least likely to buy a shoe if they think it's unpopular. In fact, a shoe with a terrible rating is still more likely to sell than a shoe with no rating. This new discovery was a completely unexpected outcome. The team had the objective of looking at new credit card data and then pivoted to start looking at ratings data.

In a typical project, this would be completely unacceptable. You don't want your teams to have a set objective and then change direction based on their own criteria. However, this is exactly the kind of empirical exploration that you want from your data science team. In fact, many data science teams try to stay away from typical objectives. They might have the open-ended notion of looking for patterns. They may just explore the data to see if anything sticks out. They want to see if there's something interesting in the data. These teams find that clearly defined objectives can often be an impediment to discovery.

When you are on a data science team, try to remember that you're doing something different from most other teams in the organization. With data science, you get the most value from your data if you focus on discovery. You should work closely with your managers to communicate the difference. Don't underestimate the challenge of trying to change their expectations. Most have spent many years focusing on meeting objectives. A team of people exploring data looking for something new might not to be easily accepted.

Taking Advantage of Serendipity

You've seen that it can be difficult to set objectives with your data science team. Still, many organizations find it hard to even imagine working without them. Objectives are everywhere: there are career objectives and learning objectives, and self-help books talk about personal objectives. These objectives guide much of what we do, but they might not be as valuable as you think.

There's been some interesting work done in this area over the last few years. It's coming from a place you might not expect: the world of machine learning and artificial intelligence. The people who are trying to get computers to display intelligent behavior are finding that much of what we know is based on unplanned discovery. We actually learn more from our wandering than we do from our set plans.

One of the best books on the topic is *Why Greatness Cannot Be Planned: The Myth of the Objective* by Ken Stanley and Joel Lehman.[1] Professor Ken Stanley runs a research group at the University of Central Florida that works on artificial intelligence. In the book he says, "objectives actually become obstacles towards more exciting achievements, like those involving discovery, creativity, invention, or innovation." This is coming from a leading computer scientist researching artificial intelligence. This isn't an inspirational quote from Deepak Chopra.

The way you should think about this is that the more you focus on objectives, the less likely you are to make interesting discoveries. Everyone on the data science team should be comfortable with creative wandering. In fact, Professor Stanley points out that the team should actually rely on pure serendipity.

Serendipity is a strange word to come out of a book on artificial intelligence. To put it simply, serendipity is when something just happens. It can't be predicted or planned. It's like when you bump into a friend on the street, and then you decide to sit down together and have a cup of coffee. It's unscheduled, unplanned, and unknown.

As strange as it may sound, a data science team has to rely on some serendipity. Sometimes a team member will see something in the data that they weren't expecting. It will look interesting or unusual. It's important for the team to follow up on that discovery. You don't want them focused on objectives at the expense of discovery.

Professor Stanley calls these **stepping stones**. These are the interesting things that eventually lead to insights. If you ignore them, you are likely to miss key discoveries.

Let's go back to our running shoe web site. The data science team has been tasked with predicting how many new sales the site should expect in the upcoming year. While looking at the data, an analyst sees something interesting. There's a slight dip in sales on Sundays over the last few weeks, as shown in Figure 14-2. If the team is completely focused on the objective, they might ignore this interesting discovery because it's too hard to imagine that the slight dip might help them create a report that predicts upcoming sales. A data science team that's focused on discovery would follow up on this interesting information.

---

[1]Kenneth O. Stanley and Joel Lehman, *Why Greatness Cannot Be Planned* (Springer Science Business Media, 2015), p. 978-3.

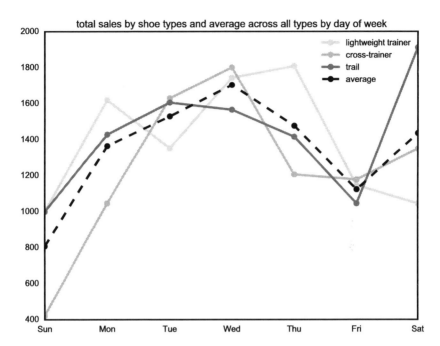

Figure 14-2. Total sales by shoe types and avererage across all types by day of week

---

Note On average, all shoe types had a slight decline in total sales dollars on Sunday, but different types peaked on different days of the week. See how to create this chart at http://ds.tips/tuc2E.

It might not lead to anything. In fact, most of these little discoveries will just be dead ends. However, a few of them will be stepping stones to something that could be very valuable in the future. The more the team explores the data, the more they will create connections to future questions.

All this language is something that you don't typically hear in organizations. Words like stepping stones, serendipity, and discovery sound more like the keywords you'd use in the trailer of a romantic comedy. Yet, these are key parts of trying to learn something new and interesting.

Canadian philosopher Marshall McLuhan once said, "I don't know who discovered water, but it wasn't a fish."[2] Much of discovery is being comfortable not knowing where your information will lead. You have to be able to pursue the unexpected. The stepping stones that you take to discovery will only be clear at the end of your path. The key is to not ignore things that look interesting just to stay true to your objectives.

Remember that data science is applying the scientific method to your data. A key part of the scientific method is making observations and asking interesting questions. Don't shortchange your exploration for the sake of short-term planning.

Adding to Organizational Knowledge

You've seen that a predictable way to make great discoveries is to allow your team to explore the data and look for interesting connections. There's also the DSLC, which forces the team to tell interesting stories every two weeks and gives the organization visibility into what the team is doing. Instead of objective planning, the organization is getting predictable delivery. Giving your team a predictable framework is a great way to keep them focused on building knowledge. It's a biweekly rhythm to share their stories. If the rest of the organization doesn't like their stories, they can always encourage them to go in a different direction.

The DSLC, sprints, and exploration all work together to deliver insights and learning. If you're working on a data science team, you should try to keep these three things in balance. The DSLC gives the team a blueprint for how to think about delivering value. The team should identify the roles and work in a cycle of questioning, researching, and discussing results.

Sprints give the organization a predictable pace. Without a sprint, the team is in danger of spending too much time preparing instead of delivering. Most of the time, your data analyst is scrubbing and preparing large data sets. A two-week sprint forces your data analyst to work in the smallest possible chunks, and encourages him or her to deliver many smaller reports instead of one large presentation.

---

[2]Quoted in a speech by Peter G. Peterson, President of the Bell & Howell Company, delivered at the Technology and World Trade Symposium, November 16-17, 1966, Chicago, Illinois, Start Page 83, Quote Page 91, Conference Sponsored by U.S. Department of Commerce and National Institute of Standards and Technology, National Bureau of Standards Miscellaneous Publication, U.S. Government Printing Office, Washington, D.C. (HathiTrust) http://hdl.handle.net/2027/uc1.b4112688?urlappend=%3Bseq=103

Finally, the organization needs to put a lot of emphasis on exploring the data. The team should have the freedom to follow up on interesting discoveries. The rest of the organization will still have visibility into the team's work, but that work could change. It could pivot based on a serendipitous discovery.

The balance between sprints and exploration helps keep the conversation alive. The team has extra freedom, and in return, the organization gets weekly feedback. If it's done well, the data science team will work closely with an organization to help employees and executives learn about the business and their customers. It's a well-balanced mix of lightweight structure and frequent discovery.

That being said, it's not an easy balance. During some sprints, the team might not be able to deliver anything interesting. Other times, the data sets will seem so large and complex that they couldn't possibly be broken down into a two-week sprint.

This DSLC framework isn't designed to solve those problems. It's just a way to shine light onto the struggle. It forces the team to think small, which encourages the organization to allow exploration.

One way to make sure that your sprint is always delivering value is to end every Storytelling Session with a clear call to action. Your audience will be very interested in adding to organizational knowledge. You can help highlight the value of this knowledge by making clear suggestions on how they can leverage this new data.

Let's go back to your running shoe web site. You saw a clear connection between whether a shoe has a rating and how well it sells. In your storytelling session, you should set up a clear visualization showing the connection between sales and ratings. The new organizational knowledge is that shoes that have no ratings are less likely to sell. Yet, that shouldn't be the title of your story. Instead, you should show how your organization could gain value from this new knowledge. You could title the visualization, "Increasing our number of rated products should increase overall sales."

With this title, you're not just saying what the organization knows. What you're doing is clearly outlining the value your team delivered. In one week, the data science team has made a suggestion on how to increase sales. There's a call to action. If you want to increase sales from the site, encourage customers to rate their products. This call to action could be directed to the rest of the organization or be redirected to the team. At the Storytelling Session, the organization might suggest that the data science team come up with an interesting story on how they can increase product ratings.

When you're working on a data science team, try to remember that your organization will view new knowledge in a very practical way. Be sure to balance the DSLC with sprints and exploration to deliver interesting stories.

These stories should have new organizational knowledge and a clear call to action. When your team has a clear call to action, you're more likely to get interesting feedback from the rest of the organization. They may ask you to follow up on your story or create new stories that give even more guidance.

Focusing on Exploring Over Routine Work

In 1999, two psychologists ran an experiment.[3] They filmed a video of six people passing a ball. They showed the video to 40 students. They asked the students to count how many times the ball passed from one person to the next. Most of the students were able to count how many times the ball had been passed. What they didn't say was that a person in a gorilla suit would walk into the middle of the screen. The gorilla stopped in the center, and then walked off camera. When asked, nearly half of the students hadn't noticed the gorilla. In fact, the participants were so convinced that it wasn't there, they had to replay the video.

The psychologists published their results and called this **perceptual blindness**. It's when people are so focused on routine tasks that they're blind to interesting events. The students watching the video were so focused on counting the passes that they didn't notice someone in a gorilla suit.

This study has been repeated dozens of times. One experiment put a small image of a dancing gorilla on a CT scan to check if radiologists would notice.[4] It turns out that 80% did not. Even people who knew about the study were only slightly more likely to spot something unexpected.

What this shows is that routine work often shuts off the part of our brain that sees unexpected events. Many people out there are doing complex routine tasks but wouldn't see a gorilla.

This can be a real danger for your data science team. Remember that much of the value in data science is exploration. You want everyone on the team to notice something interesting, yet part of the work is routine. Your data analyst still spends most of their time scrubbing the data. What you want to watch for is when your team becomes so focused on routine that they miss something unexpected.

I once worked for a company that was trying to understand why customers clicked on certain advertisements. Each ad showed a picture of a car. When the

[3]Daniel J. Simons and Christopher F. Chabris, "Gorillas in our midst: sustained inattentional blindness for dynamic events," *Perception* 28, no. 9 (1999): p. 1059-1074.
[4]Trafton Drew, Melissa L-H. Võ, and Jeremy M. Wolfe, "The Invisible Gorilla Strikes Again: Sustained Inattentional Blindness in Expert Observers." *Psychological Science* 24, no. 9 (2013): p. 1848-1853.

customer clicked on the car, the image and click-through were recorded to the cluster. The data science team created several visualizations of the data. They focused on creating real-time success rates and click-through. There were a lot of tools that helped them display this information in interesting ways. The data science team was very busy and settled into a predictable rhythm of collecting data.

In one of the storytelling meetings, the data analyst clicked down deep into a data visualization to show the detail in the data. As an example, they showed the results of an ad for a red Ford Mustang. For some reason, this ad did very well. It had a much higher click-through rate. One of the stakeholders on the team interrupted the presentation and asked why the ad was so successful. The data science team hadn't considered the question. They were so focused on getting the click-through data that they hadn't really noticed anything interesting. Their work had become routine. They gathered the data, scrubbed the data, and then uploaded it to the cluster. They hadn't asked many interesting questions. It was like a gorilla had walked into their data and they hadn't even noticed.

The data science team addressed the question in the next sprint. The research lead asked some interesting questions about the successful ad. What made this ad more successful? Was it the make of the car? Was it the type of car? Was it the color of the car? Why were the customers on the site more likely to buy this car?

After going through these questions, the team told an interesting story. It turns out that the color of the car has a slight impact on the click-through rate, as shown in Figure 14-3. This, combined with the make and the model of the car, was a likely reason that this ad was more successful. The data science team left the next Storytelling Session with a call to action. They suggested that changing more of the cars to red would improve their overall ad revenue.

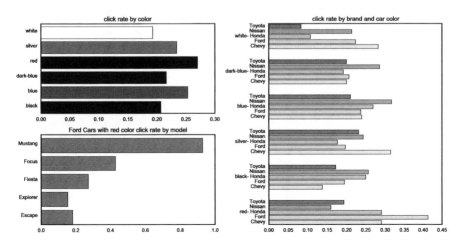

Figure 14-3. Color click rate

▓ **Note**   Across all colors, red might have a slightly higher click rate than the other colors. If you drill down by brands, only Ford has a higher-than-average click rate for the red color than other brands. For example, for Nissan, blue is the color with the highest click rate. If you further drill down by model of Fords with a red color, only Mustang has an extremely higher click rate; other models are relatively lower. See how to create this chart at http://ds.tips/wr5nU.

The team was lucky that they had a stakeholder outside the team that pointed out the gorilla in their data. Most teams will have to focus on making sure they're asking interesting questions.

Remember that your team delivers insights and not data. There's no prize for the cleanest data or the largest cluster. Still, it's easy for your team to get focused on these routine parts of data science. When that happens, your team might not notice a gorilla in their data.

Connecting Insights to Business Value

Even when your team finds something interesting, you still have to connect it to real business value. It's not easy to connect exploration to business value. If the team knew where to look, it wouldn't be exploring. Often, in data science, you don't know the business value until after you've found the insight. You have to go through the entire DSLC before you can deliver any new knowledge.

That's one of the key benefits of working in sprints. You'll deliver these insights a little bit at a time every two weeks. In each sprint, you'll build on what you know. The research lead can evaluate your insights and connect them to business value. If the teams are on the wrong path, they can pivot to something more interesting.

I once worked for a retailer who was trying to improve their worker safety. They created a Hadoop cluster that collected all of their unstructured data. The cluster had video, images, and injury reports. The data science team used this data to create a word cloud of all the organization's job injuries, and then the team presented a simple visualization of the cloud at their Storytelling Session (see Figure 14-4). As they started to tell their story, you could see everyone in the room rub their hands or cross their legs as the team described common injuries. At the end of the meeting, the data analyst said that they would use the next sprint to refine their analysis. They would create data visualizations that told a deeper story and would cover more specific injuries.

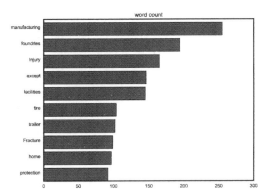

Figure 14-4. Word cloud of all the organization's job injuries

▓ **Note**   See how to create this chart at http://ds.tips/waxU8.

A stakeholder in the room asked the data science team a simple question: "why are you focused on the injuries and not on the equipment that's causing the injuries?" Everyone in the room could empathize with the people who were injured. Still, the real value is in trying to prevent future injuries. The team should have used predictive analytics to tell if the work was too dangerous.

The data science team had been focused on who was injured, which was a difficult story that needed to be told. However, the real business value to existing workers was preventing future injuries. The team needed to pivot to look at the equipment the people were using when they were injured, or at the process they were following. That was an entirely new set of data to explore.

If the team hadn't been working in sprints, they would have spent months or even longer refining and exploring the data on injuries. They would tell interesting stories, but not the ones the stakeholders wanted to hear. Instead, in the next sprint, the team focused on dangerous activities. They built on their previous data and told a whole new story about dangerous equipment and processes.

It's not unusual for data science teams to explore the data without a clear connection to business value. In fact, the Gartner Group estimates that 85% of data analytics teams work with no connection to business value.[5] Some of this is the nature of the work. It's difficult to evaluate what you don't know. The other part is making sure that you have a clear connection to the stakeholders. Your research lead will work with the business to connect the team's insights to real value.

As mentioned previously, working in sprints allows the team to quickly pivot when it finds something interesting. The stakeholders might not always know where to find business value. Instead, they're much more likely to tell you where not to go. Still, that feedback loop is essential to keeping the team on track. Knowing where not to go might eventually lead you on the right path.

The data science team should be doing interesting work. It's one of the places in the organization where you can build up real insights. Yet the team won't be immune to typical business pressures. If your data science team isn't producing real value, it won't be long before stakeholders start to question the work.

Most data science teams work much differently from the rest of the organization. If you don't quickly start to show savings, it's unlikely that you'll be around long enough to make a difference. The best way to create value is by having a tight feedback loop between the business and your data science team. The stakeholders should know what the team is working on during every sprint, and the work should be clearly connected to something that they see as valuable.

In each Storytelling Session, try to tell a simple story about what the team learned and how it will help the rest of the organization. These meetings are essential to keeping the team working and focused on interesting work.

---

[5]Ted Friedman and Kurt Schlegel, "Data and Analytics Leadership: Empowering People With Trusted Data," in Gartner Business Intelligence, Analytics & Information Management Summit, Sydney, Australia, 2016.

Summary

In this chapter, you learned how to change the mindset of the company by imagining how to work without objectives. You found out that the focus on objectives can keep your team from exploring and discovering. In Part IV, you will learn how to ask great questions. To ask great questions, you have to understand critical thinking, which you'll learn about in Chapter 15.

Asking Great Questions

Jonas Salk once said, "What people think of as the moment of discovery is really the discovery of the question." As you've learned throughout this book, one of the most important parts of data science is discovering great questions. In this part of the book, you'll find out how to do just that. To ask great questions, you have to understand critical thinking (asking the critical questions). Next, you'll find out how to encourage people to ask interesting questions and where to look for good questions.

Understanding Critical Thinking

Questions are powerful and essential to your data science team. In this chapter, you'll find out how to harness the power of questions. Then you'll learn that those interesting questions are part of critical thinking. You'll also find out about critical *reasoning* and how you can pan for gold to find the right questions.

Harnessing the Power of Questions

Imagine that you're giving a presentation to a group of coworkers. You've come up with a way to increase your company's sales—a strategy that took you weeks to prepare. In the middle of your presentation, someone interrupts you to ask a question about your assumptions: "how did you come up with your results?" How would you react to this question? In some organizations, this would be seen as confrontational and combative. Usually, these types of questions come from skeptical supervisors or someone who disagrees. Either way, it's outside the normal rhythm of a presentation.

© Doug Rose 2016
D. Rose, *Data Science*, DOI 10.1007/978-1-4842-2253-9_15

In Sidney Finkelstein's book *Why Smart Executives Fail: And What You Can Learn from Their Mistakes,*[1] he points out that many executives accept good news without question. They save their questions for bad news or if they disagree, which means that most organizations see questions as a type of disagreement. When there aren't any questions, people usually repeat the same mistakes. They're prone to groupthink and have blind spots. A lot of public failures can be traced back to crucial questions that were never asked.

As mentioned throughout this book, most organizations still focus on getting things done. They have mission statements, encourage teams to drive and deliver, and work with clearly defined goals and aggressive timelines. It's difficult to imagine an organization or meeting where everyone asks interesting questions. In many organizations, there simply isn't any time to encourage this type of questioning. However, it's important for your data science team to exist outside of this reality. Your team needs to create an environment that's open to interesting questions. The rest of your organization may live in a world of statements, but your team needs to be comfortable in a world of uncertainty, arguments, questions, and reasoning.

When you think about it, data science already gives you many of the answers. You'll have reports that show buying trends as well as terabytes of data that show product ratings. Your team needs to use these answers to ask interesting questions. It's up to you to create a setting where everyone feels comfortable questioning each other's ideas.

There are a couple things to remember to help your data science team stay on track.

First, if you have a newly formed data science team, it's unlikely that the team is good at asking the right questions. That's because they haven't had much practice. Most teams don't ask questions because good questions challenge your thinking and are not easily dismissed or ignored. They force the team to unravel what is already neatly understood, which requires a lot more work than just passive listening.

When you were in school, your teachers probably moved quickly through material because they expected you to memorize facts and read through expert advice. When you raised your hand, it was probably for a pretty simple question. It was probably something mundane, like "will this be on the test?" No one asked bolder questions like, "why are we learning this subject?" or even, "can we learn something different?"

[1]Sydney Finkelstein, *Why Smart Executives Fail: And What You Can Learn from Their Mistakes.* Penguin, 2004.

At work, you probably haven't had many opportunities to ask interesting questions. Most companies still promote people based on their ability to follow through with a corporate vision. You need to work well with your coworkers. Always asking questions isn't always the best way to get along. You need to change that view for your data science team.

The second thing to remember is that asking questions is really hard work. Most people still prefer to make simple statements. It's pretty easy to tell the world what you think. It's not so easy to *defend* what you think to someone who can ask good questions. For example, think about something that you do for yourself that's healthy. Maybe you eat certain foods or do certain exercises. Now ask yourself, how do you know it's healthy? Is it because someone told you or because of how you feel? If it's because someone told you, how do you know that person is right? Many experts disagree on what's healthy. Which experts are right?

It doesn't take long to realize that questioning can be exhausting. It takes a lot of work to deconstruct what you already believe to be true. Now imagine doing that in a group setting.

Try to remember that asking good questions is difficult to do and not always well received. Still, it's essential to your data science team. The best questions will give you new insights into your data that will help you build your organizational knowledge.

Panning for Gold

Asking interesting questions is a key part of critical thinking. So let's ask an interesting question. What is critical thinking? Most people think of critical thinking as a form of criticism. You're judging something and deciding if it is good or bad or right or wrong. So does that mean if you don't agree with someone that you're applying critical thinking? Most people would say no.

Critical thinking is not your ability to judge something. The "critical" in critical thinking is about finding the critical questions that might chip away at the foundation of the idea. It's about your ability to pick apart the conclusions that make up an accepted belief. It's not about your judgment—it's about your ability to find something that's essential.

Many organizations complain that they don't have anyone who applies critical thinking. Trying to find the critical questions isn't something you can do all the time. It's a little like running. Most people can do a little, and then with some exercise they can do a little more. Even the best athletes can't run every day.

Think about our running shoe web site. Imagine that the company gave out customer coupons and had a one-day sales event at the end of the year. At the end of the month, the data analyst ran a report that showed a 10% increase in sales, as shown in Figure 15-1. It's very easy to say that the lower prices encouraged more people to buy shoes. The higher shoe sales made up for the discounted prices and the promotions worked. More people bought shoes and the company earned greater revenue. Many teams would leave it at that.

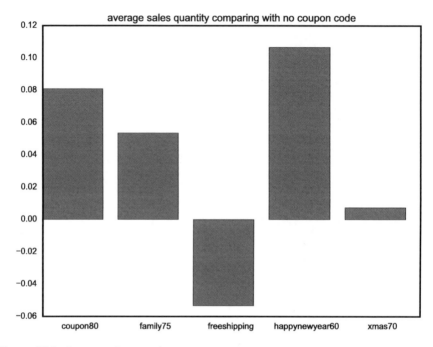

Figure 15-1. Average sales quantity

Pivoting average sales quantity by item SKU and by coupon code (including no coupon code), taking average sales quantity of each coupon code and minus the average sales quantity of no coupon code, you get how many more units sold on average when using each of the coupon code versus not using coupon code. For the coupon code with the highest discount (60%), there were on average 0.1 more unit sales than not using any. See how to create this chart at http://ds.tips/6acuV.

Here's where your data science team would want to apply their critical thinking. Remember that it's not about good or bad; it's about finding critical questions, such as:

- How do we know that the jump in revenue was related to the promotion? Maybe the same number of people would've bought shoes regardless of the promotion.

- What data would show a strong connection between the promotion and sales?

- Do the promotions work?

Everyone assumes that promotions work. That's why many companies have them. Does that mean that they work for your web site? These questions open up a whole new area for the research lead. When you just accept that promotions work, everything is easy—they worked, so let's do more promotions.

Now that the team has asked their questions, it's time for the research lead to go in a different direction and ask even more critical questions, such as:

- How do we show that these promotions work?

- Should we look at the revenue from the one-day event?

- Did customers buy things that were on sale?

- Was it simply a matter of bringing more people to the web site?

This technique is often called panning for gold. It's a reference to an early mining technique when miners would sift through sand looking for gold. The sand is all of the questions that your team asks. The research lead works with the team to find gold-nugget questions that are worth exploring. It's not easy, because determining which questions are gold nuggets is a value judgment. It is up to the research lead to determine whether the questions are interesting.

The point of panning for gold is that even though you will have a lot of throwaway questions, the few gold nuggets can change the way your organization operates. There will be a lot of sand for every nugget of gold. It takes a lot of patience to sift through that much material.

If you are the research lead for the team, try to focus on actively listening to everyone's questions. Often, their questions will be an earlier version of your question. Don't be afraid to ask big "whys." It might seem obvious to everyone that promotions work. That doesn't mean that you should ignore the question. If you're not satisfied with the answer, you may want to work with the data analysts to create reports.

Also, be sure to watch out for your own conclusions. Remember that critical thinking is about breaking down these conclusions. Make sure that you evaluate what the rest of the team is saying.

This can be really tiring work. You don't want to be in a position where you're being forced to accept a conclusion because you didn't take time to ask questions. If you're not getting to these critical questions, feel free to reschedule the meeting. Get back together when everyone feels more energized.

Focusing on Reasoning

Many of us have strong beliefs that guide us and help us understand new things. When you're working on a data science team, beliefs might strongly influence how you and other people look at the same data. That's why a key part of critical thinking is understanding the reasoning behind these beliefs. You should not just be able to describe your beliefs—you need to describe your reasoning behind those beliefs.

Reasoning is the evidence, experience, and values that support conclusions about the data. When you're working on a data science team, it's important to understand each other's reasoning. This will help the team come up with interesting questions.

Let's look at a simple statement as an example. "You should drink a lot of green tea because it's good for your health." The idea here is that you should drink a lot of green tea. The reasoning is that it's good for your health. When you apply critical thinking, you want to ask questions about the reasoning. Why is it good for your health? How do you know it's good for your health? Is it good for everyone's health? If you don't apply critical thinking, you're left with just the idea. You just accept the fact that you should drink a lot of green tea.

Now, let's go back to our running shoe web site. Imagine that the design team is exploring some feedback they received from customers. Many of the pictures on the site depict runners in peak physical condition. Your data science team is trying to determine if changing these pictures might impact sales.

Your team works with the web designers to run a few experiments. They randomly replace images of fit runners with those who are less fit and older. The team works with a data analyst to create reports to look at the difference in the data after the pictures were changed. The reports show a drop in overall sales, as shown in Figure 15-2.

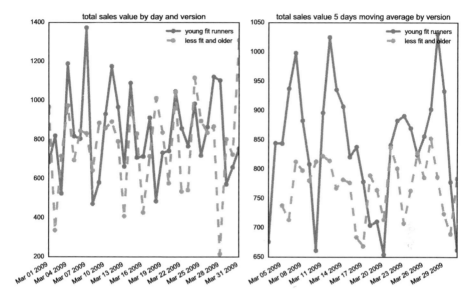

Figure 15-2. Drop in overall sales

Look at the time series and you will see that the "less fit and older" version of the page had slightly lower total sales by day. If you look at the five-days moving average, the "less fit and older" version is lower for the entire month. See how to create this chart at http://ds.tips/X3xex.

Now the team needs to talk about the results. Your project manager suggests that the drop in sales was because runners are motivated by the images. They don't want pictures that show what *they* look like. Instead, they want pictures of who they'd like to become. The drop in sales made the shoes look less effective. It blurred the message that if you buy the shoes, you will become more fit.

The data analyst disagreed and suggested that the drop in sales was because customers thought the pictures represented an ideal customer. As a result, the customers assumed that these shoes were designed for people who just started running.

To apply critical thinking, you have to look at the reasoning behind each of these statements. In these two examples, the keywords are "because" and "as a result." These words suggest that the reasoning will follow.

For the project manager, the reasoning was that customers are "motivated not by who they are but who they'd like to become." For the data analyst, the reasoning was that "customers assumed that the product was designed for people who just started running."

Now that you have the reasoning, you can start to look for critical questions. Are customers motivated to look young and fit? Did customers really believe that less fit people meant the shoes are for new runners? Who do you think has a stronger argument? More importantly, what are the weaknesses of each argument? Why would a less fit runner be considered one who just started running? You would think it would be the opposite. An older runner has usually been running for years.

There are also weaknesses in the project manager's argument. Would a customer really believe that buying a pair of running shoes would make them look younger? Does that mean that images of even younger and more fit runners would increase sales?

Now that you have the reasoning and some critical questions, you can work with the research lead to look for data and determine the most interesting questions. What's the median age of the customer who buys certain shoes? What strategies can be used to determine if they're new runners? These questions will help you gain new insights about what motivates your customer.

Reasoning can be a first step toward finding critical questions. Remember that critical thinking helps your team gain more value from their reports. You can help the research leads decide what's interesting. These interesting questions will help your team have the best insights.

Testing Your Reasoning

Think about the last time you heard someone say he or she was wrong. Not wrong about a restaurant or movie, but rather wrong about something he or she passionately believed. Can you think of any? If you can't, that's okay. It's pretty rare to see someone change his or her mind. In some organizations, it's seen as wavering or bad leadership, and it's just something you don't see very often.

A University of California physicist named Richard Muller spent years arguing against global climate change. He helped found the group Berkeley Earth. Much of his work was funded by the gas and oil industry. Later, his own research found very strong evidence of global temperature increases. He concluded that he was wrong. Humans were to blame for climate change. Muller saw the facts against him were too strong to ignore, so he changed his mind. He didn't do it in a quiet way. He wrote a long op-ed piece in the New York Times[2] that outlined his original arguments and why the counter-arguments were stronger.

---

[2]Richard A. Muller, "The Conversion of a Climate-Change Skeptic," The Opinion Pages, *The New York Times*, January 2, 2016, http://www.nytimes.com/2012/07/30/opinion/the-conversion-of-a-climate-change-skeptic.html?_r=0.

Remember that it's easy to be skeptical of someone else's ideas. What's hard is to be skeptical of your own. Think of critical thinking in two ways:

- **Strong-sense critical thinking:** When you think of critical questions about your own beliefs.

- **Weak-sense critical thinking:** When you only find critical questions to pick apart someone else's beliefs.

You probably know many more people who apply weak-sense critical thinking. They have well-thought-out arguments for what they believe and won't ever question their own beliefs. If you come up with questions, they'll do their best to defend their positions. What they won't do is build on your questions or help create new questions on their own. On your data science team, you want to apply strong-sense critical thinking. Everyone on the team should question his or her own ideas, come up with interesting questions, and explore the weak points in his or her own arguments. That's how you should apply critical thinking on your data science team.

Try to imagine what this would look like on your data science team. Let's say the running shoe web site runs a promotion and sends out a coupon to everyone who buys a product. The data science team looks at the number of people who used the coupon to make a purchase. The data shows that 8% of your customers looked at the coupon. Of that 8%, about 5% of the customers used the coupon before it expired. The data also shows that there was a jump in revenue on the day that the coupon was sent to your customers. See Figure 15-3.

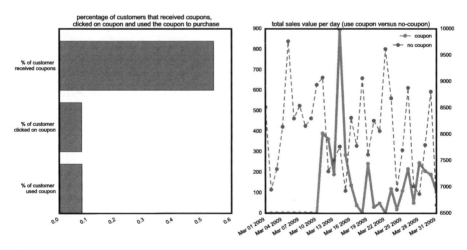

Figure 15-3. Number of people who used the coupon to make a purchase

---

The graph on the left shows that around 50% of the customers received coupons. The second bar shows that among those, only 8% of customers actually clicked on the coupon. Only 5% of those people used the coupon. The graph on the right shows a spike in sales on the day the coupon was sent to customer. The coupon did affect the "no-coupon" sales, but if you compare the actual number, coupon sales accounts only for 10% the total sales due to the fact that not many people actually clicked and used the coupon. See how to create these charts at http://ds.tips/pre6E.

Your data science team wants to see how much revenue was generated by this promotion. So let's apply some strong-sense critical thinking. You could argue that all of the new revenue that came into the site was a direct result of the promotion.

Now what are the weak parts of this argument? Maybe some of the customers who received the coupon ended up purchasing a product that was outside of the promotion. Should they be counted? Maybe you should only count the people who actually used the coupon. The problem with that is that you're not looking at the full effect of the promotion. Maybe it would be just as effective to send an e-mail asking customers why they haven't shopped in a while. This could be an interesting experiment.

Your data science team should be able to question all of these ideas. Someone on your team might feel strongly that any new revenue is the result of the promotion. That same person should also understand the weaknesses with that approach and be able to ask interesting questions, such as, "Are we fully understanding the customer if we look at the data this way?" Perhaps, the customer just needed to be reminded of your web site. If you only look at customers who actually use the coupon, it's easier to separate them into two groups: those who were motivated by the savings and those who just needed a reminder.

When your team applies strong-sense critical thinking, it should feel more like an open discussion. No one should feel like they're defending themselves. This approach is a great way for your team to ask interesting questions and, in the end, gain greater insights.

Summary

In this chapter, you learned how to harness the power of questions and that those interesting questions are part of critical thinking. You also found out what critical thinking is and how you can pan for gold to get to the great questions. Finally, you explored using reasoning while asking questions and testing your reasoning. In Chapter 16, you will learn how to encourage people to ask questions.

Encouraging Questions

In this chapter, you'll learn various ways to encourage questions. You'll find out how to run question meetings. Next, you'll explore the different types of questions and how to display and organize those questions effectively using a question board and question trees. Finally, you'll learn how to find new questions, which we'll cover in further detail in Chapter 17.

Running Question Meetings

Questioning and learning are the key differences between data science and a team just looking at data. Remember that data science is about using the scientific method to gain insights. Asking good questions is the core of this scientific method.

As we've discussed throughout this book, organizations commonly view questions as a judgment instead of a way to learn. As the leader of your data science team, you'll need to make sure that your team asks good questions. One of the best ways for you to do this is to set up a good environment to exchange ideas.

The research lead is the questioning leader who makes sure that the team asks good questions. The research lead should also focus on learning and not judging. Everyone on the team should strive for deep listening, which is a more focused way to listen to each other's ideas, and be able to push back against these ideas without feeling judged.

© Doug Rose 2016
D. Rose, *Data Science*, DOI 10.1007/978-1-4842-2253-9_16

A good way to set up this environment is to have question meetings. In these meetings, participants are encouraged to ask questions before making statements. This is sometimes called a **question first** approach. These meetings are about creating the maximum number of questions. They're focused on everyone asking their questions and listening. If you are the research lead, discourage anyone from bringing their smartphones or laptops. You want everyone focused on listening.

I once worked for an organization that was just starting out with data science. The team was in a meeting to predict how many people would participate in a medical study. The data analyst showed charts of their past studies and said he could create a data model to predict who might participate. There was a short period of silence, and then the meeting was over. A week later, the team got back together. The data analyst presented a report that showed historical data that suggested that each study might have a certain number of participants.

Now, imagine if this meeting was run with a focus on good questions. If you're the research lead, you could start the meeting by asking, "Does everybody know why we are having this meeting?" and then wait for a response. A good question leader is not afraid of short periods of silence. Don't try to answer your own questions. Give everyone in the room time to think about their answer.

Once you're satisfied that everybody understands the meeting, present the challenge. You could open the meeting by summing up the challenge with something like, "We waste a lot of money because we don't know who will show up for our medical studies." Leave the solution to the rest of the team. You could also lead with something very open-ended, such as "We need to do a better job predicting who might show up." Sit down and wait to see if anyone starts asking questions. If, after a few minutes, no one says anything, you could ask something like, "Does everyone understand why this is a challenge?"

What you're hoping to get from the team is something like, "Do some studies fill up more than others?" These types of questions allow your data analyst to come up with more complete and interesting data for the report.

What you want to avoid are quick statements that will shut down the conversation, such as, "We should do research to see how our competitors fill up their medical studies." This keeps people from coming up with their best ideas. Remember that it's the discussion that gives your team the greatest value. You want the team to feel comfortable exploring the data.

If you're the research lead, don't get too discouraged if the team has a hard time asking great questions. Most organizations still feel that it's better to be a group of people who know things (or at least speak as if they do). They like teams who speak in clear statements. This clarity is still seen as more valuable than not knowing. It might take quite a few meetings before people are comfortable asking good questions. After you have a few question meetings, you might find that a group of people who ask tough questions often gains greater insights.

Identifying Question Types

If you run an effective question meeting, you're likely to get a lot of good questions. That's great. Remember that you want your team to pan for gold and go through dozens of questions before they find a few that they want to explore further. Just like the early miners who panned for gold, you want to be able to sort out gold from sand. You'll want to know how to separate good questions from those you can leave behind.

A good way to do this is to think about different question types. Each type has its own benefits and challenges. If you're the research lead, you can help the team identify which question type leads to the most interesting insights. The two most common question types are open and closed. Each of these can be an essential or non-essential question. Some question types are easier to distinguish than others.

The first type you can identify is an open question. An **open question** has no set answer. Think about the running shoe web site. Your data science team could ask an open question such as, "Who is our ideal customer?" An open question usually requires much more discussion. These are the questions where you try to identify each other's reasoning. For example, someone on the team may say that the ideal customer is one who buys a lot of running shoes. Another person on the team might question that reasoning by suggesting that the ideal customer is one who encourages other people to buy running shoes, or is a blogger, or a runner who started a running club.

An open question isn't usually answered. Instead, it's argued. Whoever has the best reasoning usually settles the question. Your data science team will want to look for the strongest argument for who's an ideal customer, and then the data analyst will attempt to support this argument with the data.

A **closed question** is usually much more final. A closed question might be something like, "What's the average age of our runner?" These types of questions usually have some discussion. Your team might want to think about the advantages of the mean age versus the median age. They also might want to question the value of the information. What other questions could you get from this question?

If you're the research lead, make sure that the team is not asking too many of any one type of question. Asking too many open questions will make everyone spend too much time questioning and not enough time sorting through the data. Too many closed questions will result in the team spending too much time asking small, easier-to-prove questions without looking at the big picture.

Once you've identified whether your question is open or closed, you'll want to figure out whether your question is essential. An **essential question** is designed to provoke the team into deep discussion. These are questions that are typically difficult to ask in most organizations. They can be simple questions like, "Why do people buy running shoes from us?" They can also be more complex, such as, "Why do people run?"

Essential questions are usually open and have to be argued. There shouldn't be one right answer. You can also have a closed essential question. The team might ask something like, "Should we stop printing our catalog and only sell shoes through our web site?" Closed essential questions are rare.

As you may have guessed, there are also many different types of nonessential questions. A **nonessential question** is not a bad thing. You probably have to go through many nonessential questions before you start asking essential questions. A good strategy is to ask many closed, nonessential questions as a way to build up ideas and ask larger essential questions.

Let's say you want to ask why people run. Your science team might want to spend the first sprint knocking out nonessential questions. Do our customers belong to running clubs? Are most of our customers long-term runners or are they just starting out? Are there other people in their household who run? These questions might help you build a case for why your customers like running. If your team has solid reasoning for why people like to run, it will help you market your product to meet those needs.

If you're the research lead for the team, keep an eye on these different question types. They'll help you guide the discussion and sort through your highest value questions. If you know the different types, you're more likely to find the gold that will lead to your best insights.

Organizing Your Questions

If you're a fan of detective shows, you've probably seen a crime wall, used when a detective tries to figure out all the different pieces of an unsolved mystery. He or she puts up pictures and notes on a wall and tries to connect the different pieces. The board becomes a visual story. That's why you'll often see the detective sitting on the floor staring at the board, trying to pull together the story from all the little mysteries in the data.

Your data science team will have a similar challenge. They'll try to tell a story but they'll only have pieces of the puzzle. Your team can use the same technique to create a question board—a place where they can see all the questions and data. That way they can tell a larger story.

As mentioned in Chapter 13, creating a question board is a great way to display ideas and solicit questions from your team and the rest of the organization. At the very top of the board, you should put a simple identifier such as "question board" or "ask a question." The question board is a clear way to communicate and organize them in one place.

Your data science team should have dozens or even hundreds of different questions. The question board will likely be a key meeting point for the team as well as a great place for team members and stakeholders to talk about the project.

To start, place your question board next to a team member's desk or in a hallway. Open spaces aren't good for a question board. You'll want people to stand next to the board and read the questions. Another suggestion is to put the board next to an area with a lot of traffic. Ideal places are next to the water cooler, snack bar, or bathroom. It should be a place where several team members can meet and not distract other people.

Usually, the best way to organize your board is to use different color sticky notes. You'll want to organize your board from top to bottom. The sticky notes at the top of the board contain your essential questions. Use red or pink sticky notes for these questions. Below them, you can use yellow sticky notes for nonessential questions. Remember that these are questions that address smaller issues. They are usually closed questions with a correct answer. Finally, you can use white or purple sticky notes for results. These are little data points that the team discovered that might help address the question.

There are five major benefits to having a question board:

- It gives the team a shared space to help with their group discussion.

- It shows how questions are interconnected.

- It helps you organize your questions by type.

- It helps you tell a story. The question board shows the larger questions that the team might be struggling to address.

- It gives other people in the organization a place to participate. You want people outside the team to add their own questions and see your progress.

Remember that you want your team to have deep discussions. Everyone should be able to question each other's reasoning. The team should listen to each other's questions and try to come up with questions of their own. They should be focused on learning and not judging the quality of their questions.

The question board helps with this because it provides a place for people to focus their discussions. It also helps the team stand up and participate physically and come up with new ideas.

Many of your questions will be interconnected. Often, you'll have essential questions that are connected to several closed, nonessential questions. If it's on the wall, you can use string to show these connections. If it's on a whiteboard, you can just draw different colored lines. This will help your team stay organized and even prioritize their highest value questions.

As discussed in Chapter 13, the question board will invite other people outside your team to participate. You might want to leave a stack of green sticky notes next to the board. Leave a marker and a small note that invites other people to add their own questions. Sometimes these questions from outside the team tell the most interesting stories.

Creating Question Trees

Your question board will be a key part of communicating your data science story. It should include the questions that your team is working to address. It may also have little bits of data that suggest some answers. A good question board encourages other members of the organization to participate and tempts people to be part of your shared story.

One of the challenges of a question board is keeping it well organized. Since it's designed for a group discussion, you want everyone to be able to share the same information. It shouldn't have several different groups of one person's notes. If each group only has one person's ideas, that one person will be the only one to understand its meaning.

Instead, all your questions should be organized using the same system. One of the best ways to do this is by creating question trees. A question tree is a group of sticky notes all related to one essential question. You'll want to have the essential questions in the most attention-grabbing color. Usually this is either red or pink.

Let's imagine a question board for our running shoe web site. One question that your team came up with is, "Can our web site help encourage non-runners become runners?" If you're the research lead for the team, you want to put this essential question on a red sticky at the very top of the board.

Underneath that essential question, you can start adding other questions. It could be another essential question such as, "What makes people run?" It could also be a nonessential question like, "Do non-runners shop on our site?" Since this is a closed question, you could put a little data sticky next to the yellow question sticky. Maybe something like, "Data suggest that 65% of our customers don't run in a given week." You could use a pie chart like the one shown in Figure 16-1 to illustrate this point.

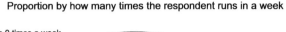

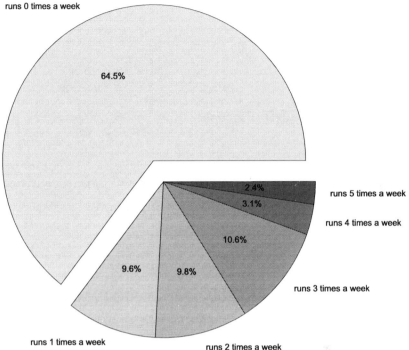

Figure 16-1. Pie chart that shows how many times per week respondents run

---

Assume that this generated data comes from a survey that the company did on its customers. The question asked, "How many times, on average, do you run per week?" When you look at the data, you see that about 65% of the respondents don't run at all. 55% of the respondents run one or more times per week. See how to create this chart at http://ds.tips/S3eve.

Someone looking at the question tree should be able to follow the thought process of the team. She should see that the lower branches of questions started with one open-ended essential question ("Can our web site help encourage non-runners become runners?") and see the team addressing that question. She should be able to follow it all the way down to different branches.

Let's say that the question, "What makes people run?" branches off in its own direction. Underneath that question is another question that says, "Do they run to relieve stress?" Underneath that is another question that says, "Can non-runners who are stressed see the benefits of running?"

With the question tree, the research lead now has a report to show progress to the rest of the organization. She could show that the data science team is working on several high-value questions simultaneously. It shouldn't be too difficult to see how gaining insight into creating customers might increase revenue.

The question trees help the research lead connect the team's work to real business value. A question board should have several question trees. At the very top of the board, there should be several red or pink essential questions. Each of these should branch down like an upside down tree into several other questions. Be sure to use different color sticky notes as discussed previously (essential questions in red or pink and nonessential questions in yellow). Sometimes open questions will branch off into different question trees, and you should end closed questions with little sticky notes that show the data.

As with any tree, you're going to want to prune your questions. This is one of the key responsibilities of the research lead. She needs to make sure that your questions lead to real business value. If she doesn't think your questions will lead to insights, she might want to pull them off the question board so the data analyst doesn't start searching for results.

▓ **Note**   The research lead usually removes questions as part of the team's question meetings. You don't want your research lead pulling questions off the board without communicating the change to the team.

One of the key things about question trees is that they actually mirror how most teams come up with new questions. Remember that data science is using the scientific method to explore your data, which means that most of your data science will be empirical. Your team will ask a few questions, gather the data, and react to that data by asking a series of questions. When you use a question tree, it reflects what the team has learned. At the same time, it shows the rest of the organization your progress.

Finding New Questions

There are two ways to help your team find better questions. One of them is someone in your organization posts good questions on your question board. These are your "gift questions." The other way is by working for good questions in the regular meetings with your data science team.

The gift questions posted on your board depend a lot on your organization. Some organizations are more collaborative. Others are more controlled. If your organization is more open, you might be flooded with questions and have to prioritize the discussion. If the organization is more guarded, you could go weeks without seeing any input.

In either case, the best way to fill a question board is to make it look inviting. A good way to do this is to create a simple sign that asks for questions, such as "Please post any questions for our data science team." Another way is to integrate the question board into interesting presentations.

Let's say someone posts the question, "What makes a perfect running shoe?" That's an open-ended, essential question. If you're the research lead, you may want to copy that question onto a red sticky note, and then put it at the top of your question tree. At the next presentation, ask the team to address some interesting gift questions and tell a story about what the data suggests. They want to tell a story about the perfect running shoe. They may talk about the combination of colors, quality, and style and back up the data with simple data visualizations. Next, let the rest of the company know where that good story came from: an interesting gift question posted on the board. This recognition encourages others to participate. Most people in organizations are always looking for better ways to participate. If you can show that their questions matter, you'll have an easier time filling your board.

Questions from outside the team are a great way to get insights. These questions are often simple and straightforward, which usually makes them very difficult to discuss. These simple questions often do the most to question our assumptions. That's why these gift questions can do a lot to keep your team focused on business value.

Unfortunately, most of your questions are not gifts. Rather, they are the result of difficult discussions with the rest of your data science team. These are the questions that are hard won and difficult to create.

If you're the research lead for the team, do your best to take advantage of your gift questions. Also, work hard with the rest of your team to find questions in the six key areas you'll see in Chapter 17. The more questions you have, the more likely your team will find key insights and tie it to real business value.

Summary

In this chapter, you learned various ways to encourage questions and how to run question meetings. Next, you found out about the different types of questions and how to display and organize those questions effectively using a question board and question trees. Finally, you briefly learned how to find new questions, which is covered in more detail in Chapter 17.

Places to Look for Questions

Getting a team to ask great questions is often not as simple as creating the right environment. Even a highly skilled data science team often needs more guidance. When you meet with your team, you'll want to focus your questions on six key areas. These areas are not the only places you'll find great questions, but they're usually a good place to start. They are questions that:

- Clarify key terms

- Root out assumptions

- Find errors

- See other causes

- Uncover misleading statistics

- Highlight missing data

These six areas all have their own style of questioning. When you ask questions that root out assumptions, they are much different from the questions about misleading statistics. Each of these areas sends the team down a different path, and each is discussed in more detail in the following sections.

These six areas are meant to serve as a guide. Not all of your questions will cover the six areas. Instead, think of it this way: If you discuss these six areas, you're bound to come up with at least a few questions. These questions will be the fuel that drives your data science team. During each sprint, your team will work to address or reprioritize questions on your question board.

© Doug Rose 2016
D. Rose, *Data Science*, DOI 10.1007/978-1-4842-2253-9_17

Clarifying Key Terms

George Carlin once joked that he put a dollar in a change machine and nothing changed.[1] It makes you wonder what type of change he was hoping for. You'll never know, because the words he used have several different meanings. Unfortunately, this is true with many words in the English language. The context in which we use words has a lot of impact on their meaning. That's why looking at key words and phrases is one of the best ways to gather interesting questions.

Data science discussions should apply critical thinking. You need to carefully look at each other's reasoning, and then question each other's reasoning so you can better understand everyone's ideas. One of the best ways to do this is to question key terms and phrases.

So let's go back to the running shoe web site. Let's say that someone on the data science team comes up with an interesting question: "do people run often because it makes them happier?" This is an open-ended essential question, which means that there probably won't be a yes or no answer. Instead, the team will have to make strong arguments that are backed up by the data.

What are some of the key terms and phrases that the data science team should question? Think about words that might have ambiguity, or multiple meanings. These are often words that are abstract and subject to some interpretation. In this case, there are couple words you might want to explore: "often" and "happier."

Think about the word "often." What does that word mean to you? The meaning usually depends on the person. For example, my wife likes going to restaurants. We try to go at least once a week. If you ask me, I'd say we go to restaurants often. If you ask her, she would say that we never go to restaurants.

Your data science team should ask questions to clarify what is meant by the word "often." You might want to ask a closed question, like "How many times does our average customer run each week?" and then put this question underneath the earlier question on your question tree.

You should also explore the word "happier." What does the word "happier" mean to you? Are your customers running because they like to run? Maybe they actually like finishing and they're happiest when they come home from their run. Maybe they don't like running but it's the only way they know how to relieve stress. In a sense, they're happier by *being* a runner.

---

[1] George Carlin, *Brain Droppings* (United States: Hyperion, 1998).

This is another area where you can ask further questions. You could go for a broad essential question like, "What makes our customers happy?" You could also try to slice happiness up into segments. Maybe ask a question like, "Do our customers run because they feel they have to?" You could even get more specific and ask, "Are our customers happiest when they're done running?"

Now you see how asking about key phrases and words can quickly produce more questions. Remember that it's the research lead's job to pan for gold as your team goes through the question meetings. Just because your data science team asks these questions doesn't mean they're obligated to follow through with results. The research lead is the one who listens to these questions and picks out the nuggets that sound the most interesting.

With this one question (Do people run often because it makes them happier?), you now have five or six other questions that might be more interesting. Your team is asking essential open-ended questions that might tie to business value.

Think about the essential, open-ended question, "What makes our customers happy?" The question might seem simple and almost trivial, but if your data science team can gain some insight, this will tie into real business value. The insights are the gold nuggets that your data science team will deliver. However, many data science teams don't go after questions like this because they feel these "key terms" are obvious. Remember that what's obvious to you may not be obvious to everyone else, so take the time to ask these questions so you can make your data science team more productive and insightful.

Rooting Out Assumptions

Another way to find good questions is to look for hidden assumptions. People make hidden assumptions all the time. There's nothing wrong with assumptions. In fact, you need them to be productive. You assume that your coworkers and doing a good job. In general, people assume that you're telling them the truth. You don't want to have to second-guess everything to get something done.

What you *do* have to watch out for is assumptions that can lead to blind spots. These assumptions keep you from exploring interesting questions, and lead your team to engage in groupthink. This is especially true on your data science team, which is why one of the best ways to get new questions is to look at underlying assumptions.

In general, assumptions have four characteristics:

- They are often hidden or unstated. Very few people start sentences by saying, "If we assume this to be true…"

- They are usually taken for granted or seen as "common sense."

- They're essential in determining your reasoning or conclusion. Your reasoning might even depend on the assumption.

- They can be deceptive. Often, flawed reasoning is hidden by a common sense assumption. Something like, "Sugar is bad for you, so artificial sweeteners must be good for you."

I once worked for an organization that examined its customer service data and realized that it had a high percentage of people calling in to order products. The data science team was tasked with trying to change this behavior because it was expensive to maintain the call center. The company wanted to encourage customers to use the web site or mobile app.

The research lead started by asking some interesting questions. Why are people calling in? What can we do to make our web site easier for customers to use? How do we get more customers to use our mobile application? Why do customers prefer to talk to a person?

A few of these questions had underlying assumptions. The research lead assumed that people called in because they didn't like the web site and that customers hadn't installed the mobile application. This reasoning may or may not be right. The important thing is not to *assume* that they're right. If the team had taken this reasoning at face value, they might have missed an opportunity to discover a key insight. In this case, most of the people who called in were professionals working on a job site and couldn't use their smartphone or access the web.

Another area where the team might look for assumptions is when they're trying to predict future behavior. Let's say our running shoe web site wants to use predictive analytics to determine which shoes would be the most successful. Maybe they found that very colorful shoes have done well in the past. They create a model that predicts a colorful shoe will be more successful, as shown in Figure 17-1.

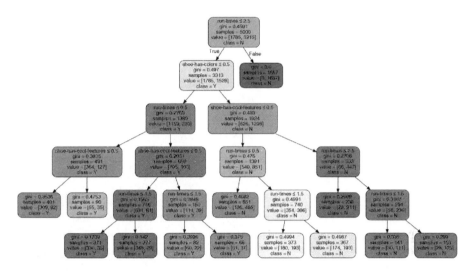

Figure 17-1. How customers decide which shoe to buy

The survey you created contains the following variables:

1. How many times does one run per week?
2. Does the shoe have cool features?
3. Does the shoe have many colors?
4. Did the respondent buy the shoes?

The fourth feature is used as a dependent variable for the decision tree to predict. See how to create this chart at http://ds.tips/fuJE3.

A good way to tackle this assumption is to try to first identify the reasoning. Here, the reasoning is that colorful running shoes have done well in the past. Once you have this reasoning, you can start to think about the assumptions.

One of the assumptions is that the running shoe was successful because it was colorful. Remember that correlation doesn't necessarily mean causation. Maybe high-quality manufacturers have been trying to make their shoes more colorful, which means that runners bought high-quality shoes that just happen to be colorful. Some of these assumptions can be highlighted with just a few careful questions. Something simple like, "Are customers buying running shoes because they're colorful of for some other reason?"

It's important to remember that assumptions are neither good nor bad. It's not bad or even critical to find out if they're all right or wrong. The key is to focus on identifying where they are. An assumption that's accepted as a fact might cause a chain reaction of flawed reasoning. Also keep in mind that an assumption isn't always an error to be corrected. It's more like an avenue to explore.

The main challenge with assumptions is that if you don't bring them to light, they have a tendency to pile up. Before you know it, your team could be working off a pile of weak assumptions. When that happens, it is difficult to make interesting discoveries.

Finding Errors

Errors in the data are not necessarily the ones that will cause the most trouble. On a data science team, a bigger problem is errors in your team's reasoning. An error in the data might be a setback or create a series of false reports. An error in reasoning, on the other hand, might send the team in an entirely different direction. The whole team may spend weeks or even months looking in the wrong places and keep the team from asking interesting questions.

You've seen how gathering good questions is like panning for gold. You can go through several bad questions before you get into something interesting. There are key phrases and language that you might want to clarify. There are also assumptions that might glue incorrect reasoning to seemingly correct conclusions. Once you peel back these assumptions and clarify the language, you should be left with the bare reasoning. In many ways, you're asking the more difficult question, "Is this reasoning correct?"

In general, there are seven dangers you want to watch out for when you're questioning someone else's reasoning:

1. **Ad hominem or personal attacks:** You might see this in your data science team if someone says, "You don't understand the data." That may be true, but it's not a constructive way to dismiss someone's questions.

2. **Question dismissal:** You don't want to dismiss an interesting question because it might lead to uncomfortable questions. You don't want your data science team to say something like, "What are the organizational politics of asking this question?"

3. **Quick consensus:** This is sometimes called ad populum reasoning. It's closely linked to groupthink and is based on the flawed reasoning that if everyone quickly agrees, they must be right.

4. **Reasoning that relies on questionable authority:** You'll sometimes see this with the classic, "I saw this chart on the Internet, so it must be right."

5. **Cyclical, or sleight-of-hand, reasoning:** Sometimes you'll see this on data science teams. You'll hear something like, "We are a data-driven company, so our data must be correct."

6. **Straw person reasoning:** This is when you intentionally distort someone else's reasons as a way to make yours look correct. You can often identify this because someone will call someone out by name. Something like, "If you accept Bill's argument that the data is terrible, we have to start from scratch."

7. **False dichotomy:** This can sometimes have a chilling effect on good questions. It's based on the idea that there are only two possible outcomes. You might hear something like, "If the data is right, it means we're all wrong."

Think of the seven dangers as a guide to keep your question meetings productive. Each of these can cause your team to have soft reasoning and shallow questions.

I once worked with a data science team on a state Department of Education testing application. They wanted to see if they could use predictive analytics to determine which students may need extra help. The team had terabytes of testing information, but they had trouble creating a working model.

In the question meeting, one team member asked, "Are these tests doing a good job assessing what a student knows?" Someone else responded, "These are the state standards and we aren't educators, so we should accept the data as correct. Besides, if it's not correct, you can't use any of the data."

These statements quickly shut down the discussion. It would've been much better if the team had explored these questions. Instead, they relied on dangerous reasoning. The state standards may have come from questionable authority. The ad hominem attack assumed that no one in the room knew enough to ask a question. Finally, there was a false dichotomy that the data was either all good or all bad.

Watch out for these dangers in your team's question meetings. It's not important to have all the answers. What is important is identifying soft reasoning. You don't want soft reasoning to take the place of interesting questions.

Challenging Evidence

As we've mentioned throughout this book, one of the best ways to find interesting questions is to look for assumptions. We've also examined the dangers of flawed reasoning. Errors in reasoning can have a chilling effect when you are looking for good questions. Once the errors in reasoning have been discovered, your data science team can think about asking critical questions about the facts that might be taken for granted.

Many organizations rely on well-accepted facts as a normal part of their daily work. These facts will be in the background while you are working with your data science team. The key is to make sure that the facts are not off-limits when asking interesting questions. In fact, your data science team might be one of the only groups in the organization interested in questioning well-established facts.

When you're on a data science team, each time you encounter a fact, you should start with three questions:

- Should we believe it?

- Is there evidence to support it? Evidence is well-established data that you can use to prove some larger fact. If there's evidence, you should ask the third question.

- How good is the evidence, and does it support the facts?

You shouldn't think of evidence as proving or disproving the facts. Instead, try to think of it as strong or weak evidence. There's strong evidence that eating too much sugar is bad for your health. There's weak evidence that honey is healthier than sugar. When you look at the evidence, all you're trying to do is decide whether or not you can depend on the facts. In this particular case, you may want to cut back on how much sugar you eat. You don't want to replace your sugar jar with honey.

When you're working on the data science team, you'll see all kinds of well-established facts. Each of them might have different sources of evidence, which can all lead to interesting questions. Some of the most common kinds of evidence are intuition, personal experiences, examples, expert opinions, analogies, and even research studies.

When you see a fact supported by evidence, don't try to look at it as a stop sign. Instead, look at the fact as a dusty hallway that probably hasn't been explored in a while. There might be something new in there to support your well-established facts. You also might find that there's no evidence at all and the fact is just an untested assumption.

I once worked for a company that had a data science team working on a set of credit card transactions. The people at the company looked at the credit card data as a way to offer promotions to bank customers. The more they knew about customers' credit card purchases, the better they were at targeting promotions.

The data science team worked with someone in the business department to try to improve the model. The team started an experiment with a new promotion on a certain brand of credit card. The stakeholder in the business department said that they shouldn't use that particular type of credit card to run the experiment because most customers who used that credit card only used it for big purchases. The data science team asked how this person knew this fact. The stakeholder said that she had been "doing this for years," and that was her intuition.

After the meeting, the data science team decided to test the manager's intuition. They created new questions on the question board. One of the questions was, "Do customers only use this type of credit card for larger purchases?" As it turns out, the manager was right. There was very strong evidence that this brand of credit card was used mostly for larger purchases, as shown in Figure 17-2. The data science team supported the manager's intuition with stronger evidence, and backed it up with transaction history and purchase prices.

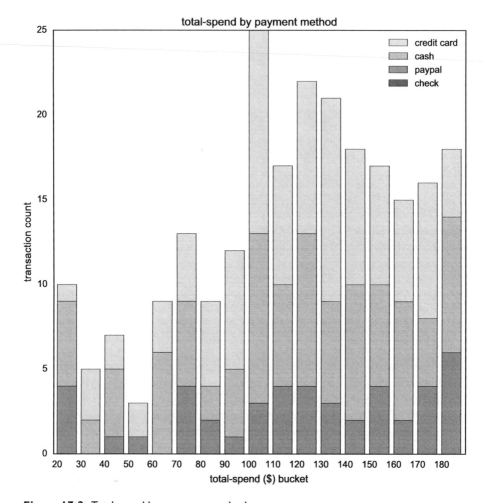

Figure 17-2. Total spend by payment meathod

If you look at the count for each transaction value bucket, as the total amount gets higher, more transactions are paid by credit card; the second category is cash. See how to create this chart at http://ds.tips/br5wR.

When you're in your question meetings, you also want to evaluate the evidence from the other team members. There's nothing wrong with intuition. Often, intuition can be the start of a great discovery. However, not everyone on the team has the same intuition. There might be some disagreement on what each team member sees in the data. When that happens, work together to look at each other's evidence. Ask questions about why one person's intuition might be more accurate than another's. Maybe they have more experience or have worked on similar projects in the past.

Just remember that facts are not always chiseled in marble. Facts can change as the evidence gets stronger or weaker. When you're working on a data science team, don't be afraid to question the evidence. Often, it will be a source of new insights.

Seeing Rival Causes

It's easy to say that correlation doesn't imply causation. It's not always easy to see it in practice. Often, you see cause and effect and there's no reason to question how they relate. Sometimes it's difficult to see that an outcome that happens *after* something is different from an outcome that happens *because* of something. You see this at work and in life.

My wife and I decided that we didn't want to buy a video game console for our house. Instead, we made a compromise with our son. We let the grandparents have a video console at their house. That way, every time we visited, our son could see the grandparents and play his new games. Each time we visited, we would buy our son a new game. It would arrive in the mail just before we left. Our son believed that we bought a new game, and then immediately flew to grandma's house so he could play. It was a very clear cause and effect.

The arrival of the video game was actually a rival cause. The game arrived and we packed our bags. Still, it wasn't the actual cause. The actual cause was that we had tickets to visit grandma, so we bought a new game.

These rival causes are not always easy to spot. There are three things to look for:

- **Whether the cause actually makes sense:** There are many rival causes out there. There's a connection between lemon imports and fewer traffic accidents. There's also a connection between ice cream and shark attacks. Still, these connections don't make any sense. Lemons don't make people better drivers and sharks don't eat ice cream. Most of your rival causes won't be that obvious. Be sure to examine the evidence on the connection between cause and effect.

- **Whether the cause is consistent with other effects:** You might find a connection between running shoe purchases and warmer weather. That means if you find a connection between running shorts and warmer weather, it's probably an actual cause. Several consistent causes make it more likely that you're looking at an actual cause.

- **Whether the event can be explained by other rival causes:** Maybe running shoe purchases go up in warmer weather because they're less expensive in the summer. If you can come up with several other rival causes, it is likely that you're not seeing the actual cause.

When you're working on a data science team, always be on the lookout for rival causes.

I once worked with a data science team for a state Department of Education. The team was creating an application to better understand student assessment data. The data showed that when students used a software program to take an assessment (instead of a written assessment), they got better scores, as shown in Figure 17-3. It made the software look extremely effective. Just the act of taking a test using a computer improved what the student knew.

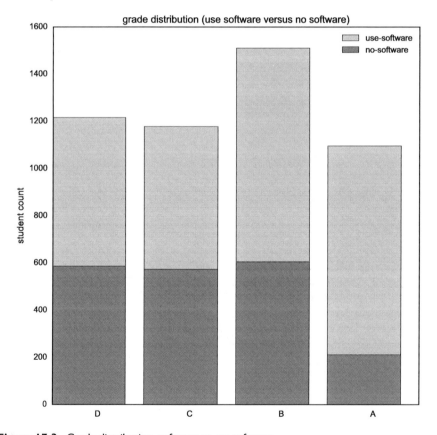

Figure 17-3. Grade distribution, software vs. no software

Looking at the total distribution, most students got a B, and fewer students got an A. However, there's almost a 50-50 chance of students receiving a C or D without software, and for students who received a B and especially an A, more of them used software. See how to create this chart at http://ds.tips/fRa5r.

It was great marketing for the software. If you used it, it would increase your scores. Practically, it didn't make much sense. The data science team didn't just blindly accept this cause and effect. Why would moving from paper tests to online tests increase student learning? Why didn't their improvement increase as they continued to use the software? This data had the hallmarks of a rival cause.

In a question meeting, the research lead brought this up. She asked the question, "Why are our students achieving higher scores?" The data science team tried to break down this question by imagining rival causes. They asked a couple interesting questions. "What other changes took place when they started using the software?" and "Were there any students who didn't have this improvement?"

It turned out that the jump in test scores had a rival cause, and there was much stronger evidence for the actual cause. It turns out that many of the schools that used the software had received state grants to improve their hardware. The classrooms each got a dozen new computers as a way to encourage the schools to use the new software. These computers allowed the students to take the test far more often, they became more familiar with the questions, and their test scores went up.

When you're working on a data science team, don't be afraid to question the connection between cause and effect. The team should be ready to create rival causes to explain certain events. If they make sense, you should investigate the connection. Some of your best questions might come from eliminating these rival causes and finding an actual cause.

Uncovering Misleading Statistics

One area where you're likely to find a lot of questions is when looking at statistics. As you've seen, statistics aren't a replacement for facts. Statistics can lie. Actually, a lot of statistics lie. At the very least, they tell their version of the truth.

When you're in a question meeting, your team should be careful to closely evaluate statistical data. They should question each other and be skeptical of statistics from outside the team.

In Chapter 4, you saw the challenge with calculating an average (the politician example). The statistical mean might give you one answer and the median might give you a different answer. Often, people prefer one or the other depending on what they'd like to see. However, there are many other ways that you can lie with statistics. Some of them are a bit more elusive. You have to listen closely to notice the sleight of hand.

One of these is statistic by inference. It's when you create a statistical connection with one story, and then you link that connection to another story. For example, let's say a study comes out that says that 20% of the time, people text and drive, as shown in Figure 17-4. A company trying to sell car insurance might come out with an ad that says, "One in five people text while driving. Make sure you have good insurance."

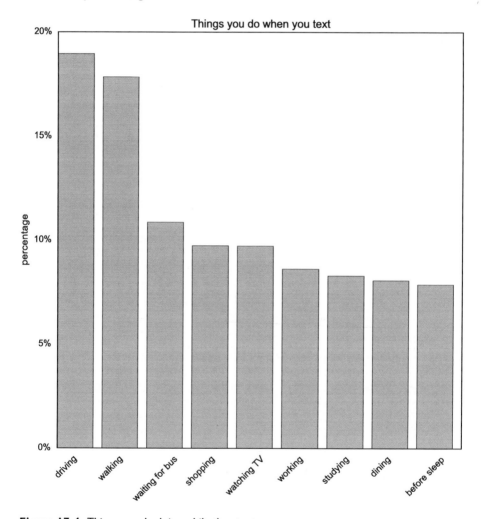

Figure 17-4. Things people doing while they text

Based on the study, almost 20% of the time, people text while driving. The second highest activity when texting is walking. See how to create this chart at `http://ds.tips/f2asP`.

Notice the sleight of hand. One statistic tells a story about text messages: about 20% of the time, people are texting and driving. The insurance story is about a group of people. About 20% of all people are doing something. Maybe there are very few drivers who are texting, but they're doing it often and that influences the data. You might text 20% of the time you're driving, but that doesn't mean that 20% of *people* are texting and driving. An average person might spend 5% of their day smoking. That doesn't mean 5% of people smoke. The story the insurance companies are trying to tell us is about safety. It tries to give the impression that everyone is distracted by texting. It makes the connection that these text messages are making it more dangerous to drive.

Let's go back to our running shoe web site. Imagine that someone on the data science team feels strongly that the data shows people are running with groups of friends, and that supports creating a new promotion. The data analyst produced a statistic that showed that 50% of your customers forwarded coupons to their friends and checked the box that identified them as friends. He suggests that as many as half of your customers run with their friends. He believes the data science team should explore whether a web site with more social interaction could boost sales. At first glance, this sounds like it makes sense. What's really happening here is that this person on your team is attempting to create statistical inference.

The best way to try to sort this out is to separate the statistic from the story. With a running shoe web site, you have two stories: one that says that customers like their friends to save money, and one that says that customers run with their friends. When you think about them as two separate stories, it's easier to see that there might be a disconnect. You may want to ask the person on your team a few questions. What's the connection between saving money and running with friends? What are some statistics that might show how many customers run with friends? Is there any connection here?

Statistical inferences can be a great tool to ask interesting questions. They're only dangerous when the team thinks of them as facts. They might be shadows of something interesting, but they shouldn't be taken as evidence.

Another area where you might see misleading statistics is when there is a suspicious omission. One place you might see this is with measures of scale. Imagine someone on your data science team uses the statistic that a promotion increases shoe sales by 5,000 orders, as shown in Figure 17-5. That might sound impressive. The only thing that's missing is a measurement of scale. You need to ask a key question. How many orders does the site typically have each month? If it's 50,000, this is a pretty good argument that you had a successful promotion. If it's five million, it probably didn't have much of an impact.

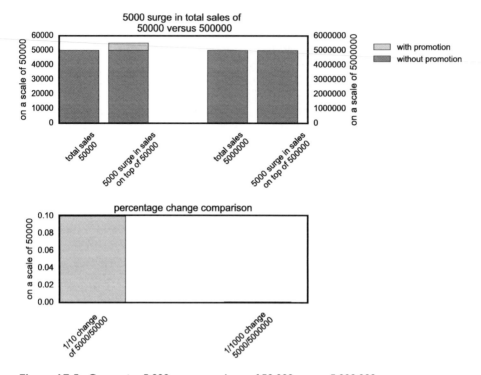

Figure 17-5. Comparing 5,000 surge on a base of 50,000 versus 5,000,000

In the top chart, you can barely notice the surge on a five million base. Similarly, if you look at the percentage change in the bottom chart, the 5,000 increase on the basis of five million is barely noticeable. See how to create this chart at http://ds.tips/tRab2.

You'll also see this with percentages. Maybe someone on the team says that red shoe sales went up by 500%, as shown in Figure 17-6. That's pretty impressive, unless of course she went from selling two orders of red shoes to twelve.

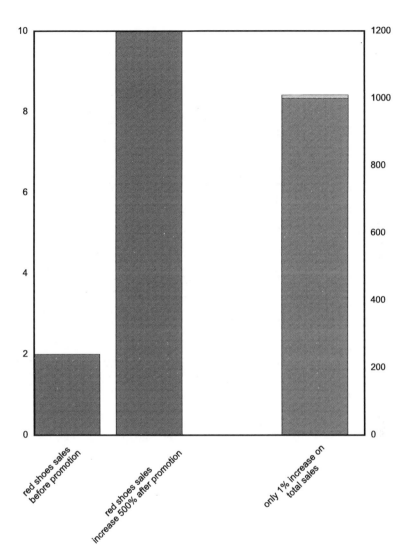

Figure 17-6. Red shoe sales went up by 500%

Red shoes sales might increase 500% after a promotion; however, you should assume that the total sales are about 1,000 per day, and that's only a 1% increase. See how to create this chart at http://ds.tips/5ugEc.

When you're working on a data science team, be careful not to accept statistics as facts. There are a few common ways to make statistics seem like they're telling an interesting story. Your team needs to ask key questions, and then ask some more interesting questions to gain better insights.

Highlighting Missing Data

One of the best ways to ask good questions is to check for missing information. A little bit of missing information can significantly change the story. Sometimes the data is incomplete. Other times, the person intentionally leaves out the information because it tells a different story. You see this a lot with advertising.

Take a very common advertisement. You may have heard the popular claim that four out of five dentists recommend sugarless gum for their patients that chew gum. That seems like a pretty strong show of support. Yet there's a bit of information that's missing. A well-placed question can change the whole story. What do dentists generally tell their patients about chewing gum? Maybe 100% of the dentists tell their patients to never chew gum. Of those patients, maybe 10% insist that they can't give up the habit.

So for that 10%, the majority of dentists say if you *insist* on chewing gum, make sure it's sugarless. As you can see, that's an entirely different story. No advertiser is going to say, "100% of dentists say don't chew gum, but for those who do, try sugarless."

On your data science team, you'll often be hunting for missing information. You'll want to watch out for information that's significant, which means it will reshape your reasoning. You can always ask for more information. The real question is what information is missing that will reshape the story. You may even end up telling a story that's much different from the original.

I once worked for an organization that was trying to figure out why more men than women were participating in their medical studies. They got a report from the labs that men were 60% more likely to participate in a medical study, as shown in Figure 17-7. The data science team was tasked with trying to figure out why this was the case. When the data science team looked at this report, they asked, "What significant information are we missing?" There were several bits of information that may have helped. They also asked, "What does it mean that they were 60% more likely to participate?" Does that mean that equal numbers of men and women applied but that a greater number of men were accepted to the study? Maybe equal numbers of men and women were accepted, but more men actually participated.

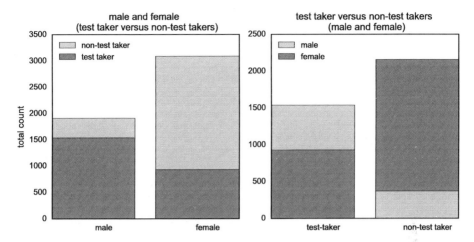

Figure 17-7. Test takers and non-test takers—male and female

You can pivot the table two ways. On the left is by male and female. About 80% of males are test takers versus only 30% of females. On the right is by test takers and non-test takers. 60% of test takers are male, while only 15% of non-test takers are male. See how to create this chart at http://ds.tips/6Wewr.

Knowing these bits of information would tell an entirely different story. One of them tells a story about more men being accepted. The other tells the story of a greater number of men showing up.

There are a few things you can try when you're looking for missing information. The first thing you should do is try to understand the reason that information might be missing. Maybe there was a space or time limitation. The person who is giving you the information might not know very much about the topic. Maybe the person has a motivation to sugarcoat the information. If any of these are the case, you might put a little extra effort into looking for missing information.

Another thing you should look out for is when the numbers are surrounded by comparative adjectives—phrases such as 60% faster, 20% greater, or 30% thinner. Often, these phrases have significant missing information from which you can derive some interesting follow-up questions, such as faster, greater, and thinner than what?

Finally, a good way to see if you have missing information is to try to take the negative view. Why does it matter that more men than women participate in the medical study? Is there a benefit to having a greater number of women participate?

It turns out that this last question helped the team find the missing information. The benefit to having more women participate is that young women were more likely to be on prescription medication, which made the study more comprehensive. They could test for a greater number of drug interactions.

This benefit was the flip side of the problem. It was more difficult for women to participate because they might be taking a prescription that wasn't allowed in the study. A better way to phrase the statistic would have been, "60% of those who were allowed to participate in a medical study were men." This tells an entirely different story.

When you're working on a data science team, try to always be diligent about looking for missing information. That significant information might hold the true story your data is telling you.

Summary

In this chapter, you learned about the six key areas on which to focus your questions when you meet with your data science team. You'll want to focus on questions that do the following:

- Clarify key terms
- Root out assumptions
- Find errors
- See other causes
- Uncover misleading statistics
- Highlight missing data

You learned that not all of your questions will cover the six areas, but if you focus on these, your team is bound to come up with at least a few questions. In Chapter 18, you'll learn how to avoid pitfalls that may occur when you're trying to ask great questions.

Avoiding Pitfalls in Asking Great Questions

In this avoiding pitfalls chapter, I'll help you discover the four reasons you might have problems asking questions and how to overcome them.

Overcoming Question Bias

Questions are at the heart of getting insights from your data. You've seen a lot of techniques to help your team ask better questions. These techniques aren't going to be very helpful unless you're comfortable with questioning. There are many different reasons why you might have trouble asking questions. These are four reasons that are very common with data science teams:

- Self-protection
- Not enough time
- Not enough experience
- Corporate culture discourages questioning

You'll see each of these in more detail in the following sections.

© Doug Rose 2016

D. Rose, *Data Science*, DOI 10.1007/978-1-4842-2253-9_18

Self-Protection

The first reason is that team members have a natural desire to protect themselves. No one wants to seem wrong or uninformed. If you're in a group with other professionals, it is especially hard to challenge someone else's answers. It takes courage to ask a good question. It makes you vulnerable, especially if you're working for an organization that places a lot of value on answers.

Asking good questions takes practice. If you are good at it, you'll find that a lot of people who seem to have answers are actually very vulnerable to questions. This helps both you and the rest of the team. If you can't address simple questions, you probably don't have a very good answer.

Not Enough Time

The second common reason is that the team just doesn't have enough time. As you've seen, questioning can be exhausting. When you're just starting out, it might seem like each question meeting gets longer and more complex. When you don't have the time to ask interesting questions, the team will simply stop asking, which makes it difficult to find any new insights.

A lot of data science teams fall into this trap. They get so focused on scrubbing the data that they don't leave any time to ask interesting questions. Often, the rest of the organization reinforces this. Doing things that are practical is seen as real work. Many organizations would prefer a busy team over an effective one. When this happens, everyone gets so focused on rowing that no one takes the time to question where your boat is going.

Remember that there is no prize for the cleanest data set. A data science team that's not delivering insights will have a difficult time creating business value.

Not Enough Experience

A third common reason is that the team simply doesn't have enough experience asking good questions. This is very common when team members come from engineering, software development, or project management. These team members may have worked their entire career becoming a person who knows the answers. It could be difficult to suppress those instincts and focus on asking questions. Team members who come from science or academia might have an easier time making the transition. That's why it might be easier to have a good mix.

When teams first start out, they have a tendency to ask a lot of leading questions. These are questions that include a version of the answer. A leading question might be something like, "I see more women are buying running shoes on our website. Do you think it's because we have more customers who are women?"

These types of questions don't really start a discussion. Usually, the only option is for the other person to agree or disagree. A better question would have been something like, "Why are women buying more shoes on our website?" Once the rest of the team starts the discussion, you can offer an opinion.

Corporate Culture Discourages Questioning

The fourth common reason is that the data science team exists in a corporate culture that discourages questioning. Social scientist Daniel Yankelovich[1] points out that most American organizations create a culture of action. When these organizations are faced with a problem, their first instinct is to rush in and create a fix. They don't want anyone sitting around asking questions. It's all hands on deck.

This type of reaction works fine in many organizations. For example, if you're working in customer service or retail, you might just focus on an immediate fix. In data science, this type of thinking will cause problems. It will keep the team from learning anything new.

One thing the data science team doesn't want to hear is a big push to finish something practical. You don't want the stakeholders to say things like, "You can ask questions once you finish uploading all the data to the cluster." That suggests that they're still thinking of your team as finishing a project and not looking for key insights.

When you're working on a data science team, watch out for an individual and organizational bias against questions. Questioning is one of the first steps toward discovery. If you skip this step, your team will have trouble learning anything new.

---

[1] Daniel Yankelovich, *The Magic of Dialogue: Transforming Conflict into Cooperation* (Simon and Schuster, 2001).

Summary

In this chapter, you learned the four reasons you might have problems asking questions and how to overcome them. In summary, the four reasons that were covered in detail in this chapter were the following:

- Self-protection
- Not enough time
- Not enough experience
- Corporate culture discourages questioning

In Part V, you will learn about the fundamental aspects of storytelling, starting with defining a story in Chapter 19.

Storytelling with Data Science

Historian Yuval Noah Harari writes in his book *Sapiens: A Brief History of Humankind*[1] that our hominid ancestors didn't become modern humans by creating tools. Instead, it was their ability to tell each other complex stories. It was our shared storytelling that was the key evolutionary step from wandering African apes into today's Homo sapiens. These stories helped us form new ideas about agriculture, justice, and religion. That's why storytelling is ingrained into our lowest levels of communication. This is especially true when communicating ideas that are new and complex. In this part, we'll talk about the fundamental aspects of storytelling. Your data science team's insights can only make a difference if you can connect the data to some larger idea. Usually the best way to do this is by weaving together a clear and interesting story. Presenting data isn't enough. For your team to be successful, your stakeholders will need to understand the meaning behind these new ideas. A good story will help bridge this gap.

---

[1] Yuval Noah Harari. *Sapiens: A Brief History of Humankind* (Random House, 2014).

Defining a Story

A coworker of mine recently got a new video camera and created a short movie about his trip to Mexico. He had software that made the video look spectacular. The opening credits looked like a movie you'd see in a theater. He had music, voiceover, and even some special effects.

We sat down together and watched his 15-minute film. About five minutes into it, I was reminded of the difference between storytelling and watching videos. He made no effort to draw me into his trip. It was just great footage of a beautiful place. I didn't connect at all to his experience. The 15 minutes went by pretty slowly. Two minutes after it was over, I couldn't tell you what I had just seen.

Many data science teams think about storytelling the same way. If you just have beautiful visualizations, then the story will tell itself. If I put up a graph that's easy to read, then the viewer will understand the meaning. In reality just like the video of Mexico, making something beautiful doesn't make it interesting. Beauty can enhance the experience, but it doesn't replace the story.

A lot of data visualization material focuses on the craft of creating charts. A data science team needs to remember that data visualization and storytelling are not the same thing. In fact, they're very different. A beautiful data visualization is like a well-designed movie set. It might stage the context, but it doesn't give you any sense of the meaning. That's why you don't watch two-hour videos of beautiful movie sets.

© Doug Rose 2016
D. Rose, *Data Science*, DOI 10.1007/978-1-4842-2253-9_19

What makes a story is not an easy thing to define. There are structural definitions. They lay out characters, struggles, and reaching an important goal. The Greek philosopher Aristotle laid out six important elements of a story. These included plot, mythos, and spectacle.[1]

These definitions are a fine place to start, but they only give you a sense of the elements of the story. They may not help your team connect with an audience. It's a little bit like trying to learn sculpture by focusing on brushes and chisels. Instead, you should think about your stories as a way to make connections.

For your data science team, try to think of a story as a way to use language and visuals to help the audience understand and connect the tale to some larger meaning.

That's one of the first things you need to think about in storytelling: how are you going to create a connection? How will you help your audience find the larger meaning?

There are a few things to keep in mind with this definition. The first is that you're using language and visuals to make connections. What you say and what you show are not in itself the story. In fact, often the visuals will get in between you and your audience.

Think about the best presentations that you've seen. Did you come away from it saying, "I didn't really understand what they said, but the chart was spectacular?" It's more likely that you said the opposite. You probably remembered the misunderstood child, or the overworked parent. The dozen or so PowerPoint slides probably faded into your distant memory.

The second part of the definition is "helping" the audience. Remember that a good story is for the benefit of the audience. There's nothing more boring than watching a data science team talk about their accomplishments. You're telling a story to help the audience connect with the material. Everything you say should be there for their benefit. That means that you shouldn't talk about the process or share the credit. Get right into helping your audience.

Finally, remember that it's all about creating a connection to help your audience find some meaning. When you've done a good job, then the audience will have found some of the meaning you were trying to communicate. Maybe they only found part of the meaning. That may have been the part where they found the closest connection. That's fine, and you can use that to build on your next story.

It's important to see that the beauty and production value of your presentation is not the same as a good story. I would've gotten a lot more out of my friend's video if he had spent less time on special effects and more time making

---

[1] S. H. Butcher, *Aristotle's Poetics*. (Macmillan, 1961).

a connection. Mexico is a beautiful country filled with colorful history and terrific stories. If he had only started there, then I would've felt like we had a shared experience. I would share the larger meaning of his trip and not just looked at the visuals.

Now that you know the general definition of a story, how do you tell a good story and keep your audience engaged with it? You'll find out in this chapter.

Spinning a Yarn

When you're giving a presentation, there are numerous ways your audience can become distracted, and it can be a challenge to keep them engaged and paying attention. There may be a clock over your head, so your audience keeps looking at the time. Nowadays, many conference rooms have glass doors and walls, which allows your audience to be distracted by people outside of the room. Trying to tell a story in a modern office is not an easy task. You need to work extra hard to focus on engaging people when they're in the room. When you start your storytelling session, you want to immediately start **spinning a yarn**.

SPINNING A YARN

This is a 19th-century term that sailors used when telling a good story. Part of being a sailor is knowing how to weave rope. Each thread needs to twist and weave together to make a strong story.

There are five key threads that your team can focus on when trying to spin a good yarn:

- Stimulate your audience's curiosity.

- Try to relate to your audience with analogies or shared experiences.

- Try not to use words like "I" or "me." Instead, use "you" or "your." You want the focus to be on the audience.

- Ask interesting questions.

- Don't be too serious. If you're funny or approachable, your audience will have an easier time accepting your ideas.

Each of these is covered in more detail in the following sections.

Stimulating Curiosity

So let's start with stimulating your audience's curiosity. Imagine that you're going into a typical meeting. The presentation slide says, "Fourth-Quarter Sales Projections." So you know there's a pretty strong upward trend at the end of the third quarter.

Imagine that same meeting, but the slide shows only the name of the presenter. The meeting starts and the presenter introduces herself. She starts by saying that the last few month's sales have been going up, but the data science team couldn't figure out why. The audience is probably asking themselves why the data science team didn't know why the sales were increasing and wondering how this story will pan out. In other words, the audience is curious. They want to see how the presenter weaves together the open question and the answer. If you keep your audience curious, they'll be patient as you tell your story.

Relating to Your Audience and Using "You"

Another "thread" you could try is to share a relatable experience. The audience needs to relate to you as a person while you're telling your story. You don't want them to think of you as a team member or a department representative. You want them to want to know what you, as a person, have to say. This will help them connect what you're saying to what they as an audience already believe.

Even though your job as a data scientist is to talk about the numbers, don't start by talking about the numbers. Instead, talk about an experience. For example, "When I first looked at this data, it reminded me of how people feel when they wait in long lines." Then go on to describe the problem of long lines and how you might lose customers.

Don't overdo it when you relate your experience to your audience. Remember that you want to minimize the use of terms such as "I" and "me." You're sharing your experience to help your audience find meaning. You're not just telling them about yourself. You're using yourself as an example for how they might approach the data.

Asking Interesting Questions

You might also want to share some of your data science teams' questions. Throughout this book, you've learned how to ask interesting questions. You can use those same questions to stimulate your audience's curiosity. If your team found the questions interesting, there's a fairly good chance that your audience might also find them interesting. Weave the question in with the experience. Make them feel that they're looking for the answer right along with you. A good question will make your audience crave the answer.

Keeping It Light

Finally, remember to not be too serious. Your audience is naturally drawn to your story when they think you're having a good time. Again, it stimulates their curiosity. They might wonder why it looks like you're having so much fun. Don't be goofy; that will chip away at your credibility. Instead, try to create a lighthearted experience. You might even want to add some humor to some of the ways your team comes up with questions. The audience is looking for your help to connect what you're saying to some overall meaning. Let them know it's a fun journey and they're more likely to get onboard.

These five threads will help you spin the yarn into a strong story. Each one of these threads will add to the overall strength of your storytelling session. You might not be able to use them all, but try to remember how they weave together to keep the audience engaged and searching for meaning.

Weaving a Story

Now that you've explored the five different threads that you use to spin the yarn, let's look into a larger topic and check out the different types of narratives you can use to rope in your audience.

A narrative is pretty much anything you might say. A television commercial is a narrative. Me saying that I waited a long time to get movie tickets is also a narrative. Not all narratives are stories, and remember that stories help an audience connect to some larger meaning. There's no larger meaning in the fact that I waited a long time to get movie tickets. I wasn't struggling to find the truth. I was simply trying to see the new *Star Wars*.

You can use different types of narratives to help your audience transition into your larger story. There are five types of narratives that are particularly helpful when you're trying to explain data science concepts to your audience:

- Anecdotes
- Case studies
- Examples
- Scenarios
- Vignettes

You find out more about each of these in the following sections.

Anecdotes

Let's start with anecdotes. An **anecdote** is a short, personal account of something that's relevant to your larger topic. The key words here are short and relevant. You want your anecdote to be long enough to be interesting, but short enough to not distract from your larger story. An anecdote is useful at the start of your Storytelling Session. For example, say you are doing a Storytelling Session on why so many of your customers are abandoning their purchase just before they check out. You might start out by telling a small anecdote about a time that you left a store without purchasing any items. You might say that it was caused by the stress of making the decision. Then, you could relay that into the larger story of why so many customers might be abandoning their purchase.

Case Studies

Another great type of narrative is a case study. A **case study** is when you relay a small problem and how it was solved. A case study is really helpful when you're trying to present a story with a possible solution, like talking about a past data science challenge and a solution that solved the problem. Let's say you wanted to use a case study to figure out why customers are abandoning their purchase. You could explain that when the web site was redesigned, there was a small drop in purchases. The design team simplified the web site and purchases went back up. The case study related to a larger story where the data science team suggested that the checkout process was too complex.

Examples

The third type of narrative is an example. **Examples** are similar to case studies, except they don't necessarily lay out a challenge and solution. They're also typically about someone else. Use examples when you're trying to justify some part of your larger story. Maybe you point out that several other companies are also trying to simplify how their customers make online purchases. Therefore, the story your audience is about to hear is not necessarily unusual or isolated to your company.

Scenarios

The fourth type of narrative is scenarios. A **scenario** is when you lay out a series of events and ask your audience to consider each outcome. Scenarios are not widely used, which is unfortunate because they're often a great way to get your audience thinking about the future. A lot of presenters think scenarios sound too childish, so if you decide to use a scenario, make sure it is not too simple.

A scenario usually works best at the start of the storytelling session. It should also be told in the third person. You don't want the scenario to sound like a personal anecdote. For example, you could relay the following scenario: Julie is on her lunch break and has five minutes left to buy the product she wants.

After three minutes, she finds the product she wants and puts it in her shopping cart. Just as she is about to check out, she sees four other products she also wants. She doesn't have the money for all five, so what is she going to do? Does she abandon the cart, thinking she'll come back later, and then forget?

Vignettes

The final type of narrative is a vignette. A **vignette** is like a little scene or a tiny movie, usually told in the third person. A good vignette will capture the audience's attention. You might want to open your storytelling session with a small vignette of your frustrated customer. Something like, "Why do they always redesign the web site? I just figured out where everything was after the last redesign."

These five narrative styles should help you increase your audience's engagement. Remember these narratives are not stories in themselves. They can help you, but they can't replace your larger story and its meaning.

Summary

In this chapter, you learned about the phrase "spin a yarn" and how to incorporate it into your storytelling using five key "threads."

- Stimulate your audience's curiosity.

- Try to relate to your audience with analogies or shared experiences.

- Try not to use words like "I" or "me." Instead, use "you" or "your." You want the focus to be on the audience.

- Ask interesting questions.

- Don't be too serious. If you're funny or approachable, your audience will have an easier time accepting your ideas.

Then you learned about the five types of narratives (anecdotes, case studies, examples, scenarios, and vignettes). You can use these when you're trying to explain data science concepts to your audience. In Chapter 20, you'll learn how to understand story structure.

Understanding Story Structure

Storytelling is not just a short description of something that happened. If you tell someone that you went to the grocery store to buy a gallon of milk, you're not telling a story. Stories have a complex and consistent structure. There needs to be a conflict and plot. In this chapter, we'll talk about the elements of a typical story. You can use these elements to weave together something that will capture your audience's imagination. It's not enough to describe the data. A complex data science story has to show why the insights matter. You'll also find that a lot of your data science stories will follow similar plots. When you see these patterns, you can structure your story in a way that will help your audience extract meaning from your team's insights.

Using Story Structure

You've seen how to take different threads and weave them together into a story. There are also different techniques that you can use to engage your audience. So now, let's look at different ways that you can bring it all together into a larger structure.

Your data science story should have three phases: beginning, middle, and ending. You should use these phases as a way to help the audience find the meaning of your story. In each of these phases, you want to do something different.

© Doug Rose 2016
D. Rose, *Data Science*, DOI 10.1007/978-1-4842-2253-9_20

In the first phase, work with your audience to set up the context. The second phase should introduce the conflict. Then, you should end the story by creating some action. Maybe you solved the conflict, or maybe the characters learned something from the struggle.

Setting Up the Context

The context is where you set up the scene and characters, introducing them and placing them in time and space. You want to set up the context as quickly as possible. Many people take too long to set up the context. You should spend just enough time to introduce the characters and place them in some setting.

For example, say your research lead is presenting in a Storytelling Session. She opens by setting up the context. She starts by saying, "We've been closely monitoring the customers who buy shoes on our web site. We can see where they live and connect that to how often they buy shoes." This sets up the context: the customers who bought running shoes, connected to where they live.

Introducing the Conflict

In the middle, start talking about the conflict. It's actually the conflict that's the most memorable part of the story. The research lead from the previous example might say, "The customers who live in an urban area are more likely to buy running shoes. In fact, the more densely populated the area, the more often they buy running shoes. We thought this was strange. As runners ourselves, we don't really like running in densely populated areas. There are too many large vehicles and too much traffic. So we decided to run a few experiments."

The conflict is where you draw in your audience. They already might be thinking to themselves that this is unexpected. The research lead uses a personal anecdote to stimulate their curiosity, and they might be even coming up with their own theories. Maybe they think it's because the customers are younger, or that they live closer to large parks?

Next, the research lead wants to create some action. This is where she addresses the conflict by explaining the solution to the conflict, where she looked for the data, and what she found. She should talk a little bit about the action taken, but at the same time not explain too much about the details. She might say, "We ran an experiment to look at their age. These customers tend to be younger, but once we adjusted for that, there was still a pretty big discrepancy. We also looked at some maps where we had a lot of active customers. We wanted to see if there were more runners' paths. It turned out that generally there were nicer paths outside of the city."

Now, the research lead draws the audience into the struggle. She does not want to spend too much time talking about all the experiments. At the same time, she still wants to stimulate their curiosity and maybe even build up some expectation.

▓ **Note**   You'll find out more about how to present the conflict later in this chapter.

Ending the Story

To close out the story, she says something like, "It turns out that the strongest connection we found was that if customers lived within three miles of a gym, they were more likely to buy running shoes." This is what you see in figure 20-1.

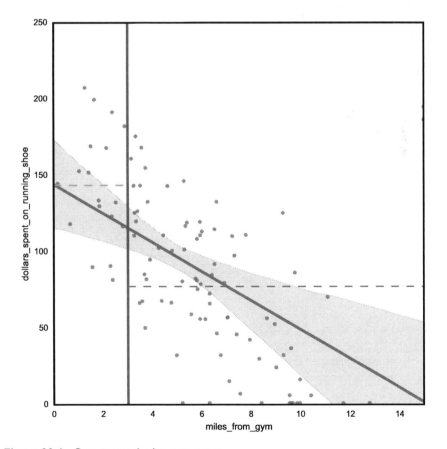

Figure 20-1. Customers who live near a gym

The orange dash line shows average dollars spent on running shoes of customers living less than three miles from a gym, and the gray dash line shows average dollars spent on running shoes by those living more than three miles from a gym. There's a clear negative correlation between these two variables. See how to create this chart at http://ds.tips/pUhe3.

She closes with a small vignette by saying, "So imagine our customer who lives close to the gym. He is running indoors and is always looking for new ways to stay in shape. Just being next to the gym is enough to make him buy more running shoes."

She closes out the story by introducing the new insight. Maybe she even asks the audience if they have any questions, and uses those to create a series of new questions to ask for the next Storytelling Session.

Giving your story some structure helps your audience gain meaning from your story. Remember that you want to put the most effort into the middle of your story. The audience is most likely to remember the conflict. Then you can close out with an action item and even get more questions for deeper insights.

Introducing the Plot

In the previous section, you learned that there are basically three phases in a story (context, conflict, and ending). In between the context and conflict, you want to include a plot. The characters and the plot are what make the story memorable. For example, in Shakespeare's *Romeo and Juliet*, Romeo and Juliet are the characters and their fobidden love is part of the plot. Their heartbreak and death is the final piece of the plot (sorry about the spoiler).

In data science, the plot is the bulk of your storytelling session. It's what the data says and your interpretation of what that data means. The plot of your story doesn't have to be original to be interesting. It's how you weave together the plot and characters in context that makes your story interesting.

In Christopher Booker's *The Seven Basic Plots*,[1] he argues that all stories just have a few plots. He says that human beings have very similar psychological needs when listening to stories. Not everyone agrees this is true for literature, but it's almost certainly the case for data science stories. Your audience will only be looking for a few different types of plots.

---

[1]Christopher Booker, *The Seven Basic Plots: Why We Tell Stories* (A&C Black, 2004).

Booker's seven plots are:

- Rags to riches
- Overcoming the monster
- The quest
- Voyage and return
- Comedy
- Tragedy and rebirth

Keep these plots in mind as you tell your data science story. These plots help define exactly what you're trying to communicate. It's pretty unlikely that a data comedy would be well received by your audience, but the other six plots just need a little tweaking to apply to your storytelling session.

Rags to Riches

One of the most common stories is "rags to riches." Almost every organization is interested in an insight they can exploit to generate new revenue. Maybe your team has an idea for a new product. Maybe you've figured out a way to scale the product you already have. When you're telling this type of story, think of how you can tell the story in a way that's consistent with a rags-to-riches plot. Describe where the organization is now and lay a clear path to the future riches. Remember to clearly lay out the plot to help clarify your story and help the audience find that meaning.

Overcoming the Monster and the Quest

Two other common plots for your data science story are "overcoming the monster" and "the quest." Many organizations try to use data to address dangerous challenges. Maybe your product has a new competitor, and your data shows a serious drop in sales. Focus your plot around overcoming this monster (the competitor) with some clever tricks that you learned from your data. Encourage your audience to embark on a quest. This is especially true if you're trying to sell your audience on doing something different, like trying a new business venture, or introducing a new product. This plot romanticizes your destination.

Voyage and Return

A less common plot is the "voyage and return," which is sometimes called a postmortem. For example, you started a new project and decided that it wasn't worth pursuing. Now the team needs to go through the data and determine whether there are any lessons to learn or if there is any value in the experience. As data science becomes more popular, you'll see more of these plots. The organization is going to want to learn lessons from these failures. So, in the future, you may see more voyage and return Storytelling Sessions.

Tragedy and Rebirth

One story you don't hear very often is a data science tragedy. Most organizations prefer to sweep their tragedies under the rug. You might see data science tragedies more with government projects for which there is an audience who is very interested in understanding the full scope of the tragedy. A data science tragedy story would be a full analysis of everything that went wrong. It wouldn't focus on lessons as a postmortem does; it would just focus on understanding the sorry story in its entirety.

Finally, you might have a story about rebirth. Sometimes companies look at the data and decide that their current line of business is unsustainable. Even large companies like IBM might decide to completely change their business. IBM moved from a low-margin PC and hardware sales model into a higher-margin service and consulting business. They did this because the current CEO was able to tell a compelling story. It was a high-technology version of the Frog Prince story.

Presenting Conflict

Remember that a story is spinning a tale with visuals to help connect the audience to some meaning. People often think that you connect to others through tales about success and achievements. That's why a lot of business meetings start out with some new achievement or goal. The audience might applaud, but they're not really connecting to any meaning. It's actually the struggle, or conflict, that helps people find meaning and connects your audience to a story.

When you're presenting a Storytelling Session, you probably have an audience that spends their entire day filtering information. This is especially true if you work with high-level stakeholders. They go through hundreds of messages and probably look through dozens of reports a day. To be an effective data science team, you have to communicate with your audience in a different way.

You've seen how your data can tell a story. In fact, you've seen how the same data can tell several different stories. Your challenge is to take lifeless data and reverse engineer it so that it contains some of the humanity of the people who created it. You want to communicate a struggle and create a plot for a data science story.

So what does this look like? Imagine that you're working on a data science team for a large credit card processing company. Your team finds out that credit card customers change their spending patterns just before they have trouble paying off their account. These customers increase the amount they use on their credit card just before they get into financial trouble. Your data science team sees this pattern in hundreds of thousands of credit cards.

Your research lead could present this information in several different ways. Maybe she presents a simple line diagram that shows an uptick in spending just before a customer gets in trouble, or tells a tale about the hundreds of thousands of customers. Both of these scenarios present the information, but probably won't connect your audience with the data. The best way to present the information is to tell an interesting story that includes a real plot with struggle and conflict.

Start by creating a character with a real name. This shouldn't be an actual customer's name, but it can be a character based on most of the common traits of your customers. Your research lead can start her storytelling session by saying something like, "Our data science team wants to talk about one of our customers today. Let's call him Allen. In two months, Allen is not going to be able to pay his credit card bill. He's been a customer with us for about six years. Over the last two months he's spent up to his credit limit. That's pretty unusual for Allen. He's 48 years old and is using his credit card to pay for groceries and transportation. Usually, he only uses his credit card for airplane tickets and hotel bills. We know Allen is not going to be able to pay his bill. Now what do we do about it?"

By presenting the data this way, the research lead has combined the data on hundreds of thousands of customers and created a plot with a real human struggle. Your audience should be able to connect with this data in a much more interesting way. Maybe they're thinking about whether they have an obligation to Allen. Should they send him a letter or call him?

It's not only important to give the character a name, but you also want to fill in some of the details about the character's life. The audience found out that Allen is 48 years old. He's been a customer for six years. These details help enhance the struggle and build out the plot.

The audience will have a much easier time connecting with the meaning behind the data. It's far easier to talk about what to do about Allen than it is to talk about the hundreds of thousands of customers he represents. There's a real plot with a real struggle that's enhanced with a few details.

Even though Allen doesn't exist, he becomes real for the purpose of the story. He can help show the conflict in a way that the data on its own may not. What was just static data became a story with a real struggle, details, and a plot.

Summary

In this chapter, you found out how to bring all the elements of your data science story into a larger structure. You learned how to work with your audience to set up the context, introduce the conflict, and create some action. Then you examined the different types of plots that you should include between the context and conflict, as well as how to present conflict. In Chapter 21, you will learn how to define story details.

Defining Story Details

In Chapter 20, you saw that people are more likely to connect with a story if there is a plot and conflict. The struggle draws people into your story. After you've established your plot and conflict, you want to keep your story dynamic. A good way to do that is to attach details to parts of your story. The details are like little mental sticky notes that help your audience remember the larger plot and struggle. It helps them create a mental image as they listen.

I once worked for an organization that was trying to use data science to get people to participate in their medical studies. It turns out that a lot of people are afraid of needles—specifically, the needles used in blood tests. It turns out that there's a cross-section of people who are afraid of needles and blood. This cross-section is a lot of people.

Not being a fan of either of these, I can certainly understand how this impacts a medical study. If needles are involved, you lose a lot of people. If needles and blood are involved, you lose an even larger group of people. This fear of needles and blood left the organization in a bit of a bind. They needed people who weren't normally interested in studies to start participating.

The data science team asked some good questions and created several reports. These reports led to an interesting story. They found that if someone participated and had a really good experience, they were more likely to participate in a future study. So if someone who didn't like needles had a positive experience in a needle-free study, that person might participate in a future study that included needles.

© Doug Rose 2016
D. Rose, *Data Science*, DOI 10.1007/978-1-4842-2253-9_21

The data science team wanted to tell that story. The research lead decided that she wanted to use a real participant and just change her name. All participants had filled out an in-depth application and were evaluated by a nurse who also filled in some information. Our research lead used some of this information as part of her story. These applications were a treasure trove of different details that provided both the nurses' observations and the participants'.

The research lead started out with a small anecdote. She said, "When I was a nurse, I could always tell who was afraid of needles. They always crossed their arms in a certain way. They grabbed both of their elbows as a way to protect themselves from the poke in the arm. There are a lot of people out there like that, and we need them to participate in our medical studies. So I'm going to tell you a little bit about someone I found in one of our reports.

"Let's call her Tracy. She participated in one of our medical studies for a drug that is being developed to help people sleep. On the first day of the study, she showed up with her own pillow. She must've been optimistic about how well it would work. She was hoping that this new pill would help her since she had some trouble sleeping during periods of high stress.

It turned out that Tracy was one of the participants who didn't get any benefit from the drug. When she left, she told the nurse that her father was a doctor, so she felt some obligation to participate in medicine. She said she could never be a doctor because she was scared of both blood and needles. A few months later, she decided to participate in a flu vaccine trial. The study required needles for the vaccination and for later blood tests. So why did Tracy decide to participate?"

The research lead ended her story by describing a call to action. The data science team thought that getting people involved in studies without needles was the best way to increase the potential number of participants.

Now, think about the story you just heard. What are some of the things that you remember? Do you remember the name of the participant? Do you remember why she participated in the first study? You might, but chances are you remember the details. The little tidbits that help you create a mental image. You probably remember that she brought a pillow or that her dad was a doctor.

These details help you carry your story from beginning to end. They create snapshots that help your audience get the whole picture. When you tell your data science story, try to use these little facts to add life to your story. They help your audience connect to the plot and the struggle.

Reporting Isn't Telling

Business presentations are boring. They're not really structured to be interesting. They're designed to communicate your status. They're like a verbal "reply all" to your stakeholders. That's usually fine for typical status meetings, but you want your data science team doing something different.

Remember that data science is about applying the scientific method to your data. Your team will explore the data, go through questions, look for key insights, and explain a lot of different outcomes. Most of your team's challenges will be around explaining the data. You need to break the data down and explain what it means.

I once worked with a data science team that was focused on delivering promotions to credit card customers. The team asked a lot of interesting questions. Some of them were about their customers' purchasing habits. One of the questions led to a key insight. The team wondered if customers were accepting promotions in bunches (more than one promotion at a time). It turned out that that question led to some really interesting reports, which showed that if a customer accepted a promotion, he or she would be more likely to accept the next bunch of promotions.

The data science team wanted to present this insight at one of their data storytelling meetings. The research lead came up with a presentation, but it wasn't a story. She just wanted to explain the findings, which were still exploratory. The presentation didn't make any attempt to explain what the findings meant. The team just wanted to point out that customers were more likely to accept promotions in bunches. Then they left it up to the room to decide what to do with that information.

I reminded the research lead that the Storytelling Session is not the right place to just present information. She needed to weave together an interesting story so the room would be engaged and connect with the meaning.

I asked the research lead why she thought customers were accepting promotions in bunches. She said that the data suggested that most consumers got money in batches, which makes them spend more during their up-times and less during downtimes. Customers were also more likely to accept promotions during certain phases of their lives.

We both agreed that this was a much better way to convey this information. Everyone in the room would've had the shared experience. They all had less or more money to spend during different times of their lives. Why not use this shared experience to convey the information about how the customers were accepting promotions? The research lead came up with a new presentation. She created a story called "The impact of promotions during different times of people's lives."

She started her presentation by relating an anecdote from her past. She said that when she was in college she had a roommate. They used to receive coupons in the mail that offered two-for-one meals. After class, her roommate would come home and ask what was in the mail so they could see where they were going for dinner that night. She said after four years of college, they ended up having the same taste in food.

A few people laughed—and they had already started thinking about how people use business promotions during different times of their lives. It was a much easier transition for the research lead to present the rest of her story. She described how most customers accept promotions at different rates at different times of their lives. That instead of a steady stream, customers in similar situations would accept promotions in bunches.

The anecdote combined with the story did a lot to encourage the group to participate. Many audience members became very curious and asked questions such as, "Does that mean that a promotion might be more effective if someone feels like they're in a vulnerable time in their life?" Another member of the audience asked, "How far is the team from predicting when a customer might be entering a phase when they would accept a lot of promotions?"

If she had just given a typical presentation meeting, she would never have had this level of engagement. The story drew them in and helped them connect the data to their own experiences. The audience thought of times when they might've been more likely to accept a promotion. Then they were able to ask interesting questions and build on this insight.

Knowing Your Audience

When you're telling a story, one of the biggest challenges is knowing your audience. Each person in your audience has his or her own view of the world and will be listening to your story with a whole host of assumptions and beliefs. Your audience is there because they want to hear something from you. They may not know what it is yet, but they will respond to it when they hear it.

I once worked for a large political campaign. The campaign was trying to use technology to better understand their audience. A few months into the campaign, the candidate was speaking to an audience that contained people who'd lost many jobs over several decades. He stood in the skeleton of an old industrial building and talked about new job training. He told a story about how each person could benefit from high-tech skills. The audience clapped but didn't really connect to the story.

A few days later, the candidate's opponent went to a similar setting. He told a speech in an old abandoned warehouse next to a lazy brown river. He started out the story by saying, "I know many of you are uncertain. You're not sure that your way of life has a future." Then he went through a short story about how to preserve the things that are important. At the end of the story, the glassy-eyed audience clapped until the candidate left the stage.

The first candidate obviously did not understand his audience. The audience didn't want to go back to learn how to become accountants. They just wanted things to go back to the way things were. Once the second candidate spoke to them on that level, they were able to connect with the story.

Political campaigns are certainly different from data science teams; however, the principle still applies. The more you know your audience, the more effective you'll be in telling your story.

One of the best ways to get to know your audience is a technique called **warming up the room**. This is when you walk around and chitchat with some of the people from your audience. Some of them will tell you what they're looking for directly. You might hear comments such as, "I am curious to see how this connects to what I'm working on." Then you can ask her, "What are you working on?" If something like this happens, you might want to adapt your stories in real time to meet your audience's expectations.

You can typically think of your audience as being divided into five different groups:

- **Observer**: The observer is someone who just showed up because the meeting was on his or her calendar. This person is not very aware of the topic and has minimal expectations. There's not much you can do for this audience group. Just try to keep your stories interesting and limit your acronyms or technical jargon.

- **Gatekeeper**: The gatekeeper is in the audience to see how your story might impact his or her work. This is a good example of the type of people you'll find when you warm up the room. If you use an example in your story, try to directly connect it with someone who is part of that person's department. This will help the person make an explicit connection.

- **Manager**: A manager in your audience is interested in interdependencies. Again, you can use examples to show departments interacting. You can also create explicit action items at the end of your story. This audience group is typically the one that asks the most follow-up questions.

- **Expert**: The experts in your audience will always push for greater detail. If you're not careful, the experts might derail your story and make it less interesting. If they do that, be sure to explain the new details so the rest of the audience can continue to participate.

- **Executive**: The executives in your audience want to glean answers to larger questions. It's always a good sign if, at the end of your story, the executive asks a question such as, "How do you see the impact of this on the rest of the organization?" If you have executives in the audience, don't show too many slides. If they're staring at your slides, they're not listening to what you're saying.

If you work hard to identify your audience and understand what they're looking for, there's a better chance they'll connect with your story. If you break your audience into these groups, you'll have an even better chance of meeting their expectations.

Believing What You Say

John Steinbeck once said, "If a story is not about the hearer he will not listen . . . A great lasting story is about everyone or it will not last. The strange and foreign is not interesting—only the deeply personal and familiar."[1]

When you're telling a story, the most persuasive ability you have is your own interest in the topic. Your audience will always be checking to see if you believe what you're saying. They'll have an easier time connecting with your story if they sense that you are committed to telling it.

When I was in law school, I took a course on litigation. This class was on how to connect to juries about what happened to your client. The jury would always be curious about your client's story. How did they get there? Why are they being judged?

Our professor had been speaking to juries for decades. His long white beard gave him an almost Jedi-like appearance. He gave us some simple advice. He said that when telling a story, try to not make it sound extraordinary. Don't try to create some far-fetched story about what happened. Instead, focus on what you know. Tell a good story about something ordinary, because what you know is the only thing that you can truthfully present. A jury can sense when you don't believe what you're saying. Say what you believe, even if it's simple and ordinary. Just say it with clarity and passion.

The same thing holds true when you're trying to connect with your audience. It'll be very difficult for you to tell your story if you don't believe that it's interesting. It's very difficult to fake passion.

---

[1]John Steinbeck, *East of Eden* (New York, NY: Penguin Books, 1986).

If someone's passionate about her topic, she can almost always talk about it in a way that makes it interesting. I once heard a presentation that was about the free airline magazines you get while flying. The presenter was so passionate about the topic that it drew people in. I've also heard presentations about international political upheavals. You could tell that the presenter wasn't interested in the topic and it was described in a way that was technical and uninteresting.

You see this a lot with data visualizations. Some research leads think that a good data visualization can add "pop" to an uninteresting story. Even the most beautiful graphics will not make your story more interesting if you don't find it interesting yourself. Your audience will get the interest from you, not from whatever you put up on the screen.

There are a few things you can do to help present a story in an interesting way:

- Make sure the topic is interesting. If you can't present something interesting, don't present anything at all. If you're not interested in how many people purchased red running shoes, you're not going to be able to tell an interesting story about it. Look for something interesting in your story. If you can't find it, you shouldn't tell the story.

- Connect yourself to the story. Tell your audience why you found it interesting. Maybe tell a story about how you were traveling to the American Southwest. You noticed that people were wearing colors that you didn't see in other parts of the country. So when you came back, you wanted to see if there was a way to better cater to this group of customers. Talk about the steps you took to get there.

- Sound like a real person. Many organizations put a lot of pressure on their employees to be efficient and superhuman, almost like an army of Vulcans who only focus on the performance aspects of their job. They shun passion and emotion. That doesn't work well for storytelling. People have an easier time connecting with someone when they're vulnerable and can laugh at themselves.

- Share your sincerely felt emotions. You don't want your Storytelling Session to have the feeling of a support group, but sharing a feeling will help your audience believe what you're saying.

Remember that you are the most important thing that you bring to your presentation. Beautiful charts, clever anecdotes, and piles of data won't make up for the passion that you bring to the topic. Even the most extraordinary data will seem boring if you can't explain it in an interesting way. The key is to make sure that you believe that the story is interesting. If you can't convince yourself, you can't convince your audience.

Summary

In this chapter, you learned that instead of just reporting the data, you need to do something different. It's important for you to be familiar with your audience and know the different types of people in your audience (observer, gatekeeper, manager, expert, and executive). You also learned to believe what you say so that you can present your story in an interesting way. In Chapter 22, you find out how to humanize your story to get people to identify with what you're telling them.

Humanizing Your Story

We have spent a lot of time talking about stories. You've seen the difference between a simple tale and a story that has a plot, characters, and conflict. You've also seen how to present a story in a way that engages your audience. Still, this is a book about data science. It's about using the scientific method to better understand your data. Ultimately, you'll have the data and want the audience to connect with it in a meaningful way. Then you'll get the audience to create value by having them take some action.

You've seen how to create the story first, because that's how you'll present your data. Now you need to reverse engineer the swirl of your data and reflect the flawed, emotional, and unpredictable human beings who created it. That's the main challenge and what separates data science teams from data analysts. Your job as a data science team is to reveal the humanity behind the numbers, which is why you shouldn't just communicate the information in the language of numbers.

Think of it this way. You're at the airport and you find a misplaced cell phone on an empty seat. This person has left the phone unlocked and you have access to all of their data. How do you find the owner of the phone? What data do you need from this phone to find the owner?

© Doug Rose 2016

D. Rose, *Data Science*, DOI 10.1007/978-1-4842-2253-9_22

Chances are, you wouldn't start by analyzing the relationship between the expense of the phone and the likelihood that someone will return it (the statistical model). In addition, if you retold the story, you probably wouldn't use the language of data and statistics; for example, "I found a smartphone but I left it because there was a high probability that someone would return." Instead, you would start by recreating the story in a more "human" way. Maybe this person was waiting for a flight and ran off to get something to eat. You look at the recent calls on the smartphone to see if the owner called someone just before he left. You know that people often call their husbands, wives, girlfriends, or boyfriends just before getting on a flight. Maybe you look in the calendar to see if there's flight information.

You aren't thinking of the smartphone as a data repository; you're thinking of it as a cherished device with a person's photos, videos, and contacts—something that someone would miss. Ultimately, you would work with the data because that's where you'd find the smartphone's phone numbers, calendars, and contact information. During this thought process, you start and end with the story, and the data is just the vehicle in between—hopefully, at the end of the story the person is reunited with their cell phone.

In Paul Smith's book *Lead with a Story*,[1] he describes how the CEO of Procter & Gamble would come to presentations and sit with his back facing the slides. Smith describes giving a presentation to the CEO, who didn't once turn around to look at the data. After the presentation, he realized that this wasn't by accident. CEOs in large companies see data all the time. They know the data is the vehicle and that the story the presenter tells has all the value. That's why when you're doing a Storytelling Session, you don't want to put too much emphasis on the data. The data won't be worth anything unless it connects with the audience, and the data on its own can't do that. It's the story that you tell about the data that helps the audience connect with the meaning.

During your presentation, you want your audience to have their pens down and their laptops closed. You want them to look at you and only glance occasionally at the data that you're presenting. If they spend too much time looking at charts or graphs, then chances are they're thinking about something else. Only after the story connects your audience to the data and gives it some meaning will your audience be spurred into taking action.

---

[1]Paul Smith, *Lead with a Story: A Guide to Crafting Business Narratives That Captivate, Convince, and Inspire.* AMACOM Div American Mgmt Assn, 2012.

Introducing Visuals

There are a number of great books and courses on data visualization. In Chapter 8, I recommended *Storytelling with Data* by Cole Nussbaumer Knaflic[2] and *The Visual Display of Quantitative Information* by Edward Tufte.[3] Both of these books have a very tactical view of storytelling. They talk about how great visuals can create great stories. They imply that data visualization goes hand in hand with great storytelling. You see this throughout both the books.

Storytelling with Data covers six lessons. The earlier lessons cover things like choosing a display and eliminating chart clutter, and the last lesson is about telling a story. You should think of these lessons in reverse, with the storytelling lesson appearing before the others. Your data science team needs to make sense of the data and tell a story to help the audience find some meaning before considering charts and graphs.

These books and courses are terrific, but they overstate the importance of data visualization. Charts and reports can certainly help tell your story, but it's the quality of the story that connects your audience to a larger meaning. The visualization is just a small part of this effort.

Your data visualizations can harm your story as much as help it. Too many visuals can be a distraction because every time you show a new image, it takes time for people to process the change. Be conservative with how much data you present when engaging your audience. The books give you a lot of great ideas on how to show the maximum amount of data with the minimum amount of clutter. Hopefully, your audience only needs a moment to glance at the visualization to understand what it says. Then they can go back to listening to the larger story.

The most important thing to remember is that data visualization is the sauce and not the meal. A really interesting story doesn't need good visualizations. At the same time, even the best data visualization doesn't cover up a boring story.

---

[2]Cole Nussbaumer Knaflic, *Storytelling with Data: A Data Visualization Guide for Business Professionals* (John Wiley & Sons, 2015).
[3]Edward R. Tufte, *The Visual Display of Quantitative Information, 2nd Edition* (United States: Graphics Press USA, 2001).

If you're the research lead for the team, there are a few things that you can do to simplify your visuals so they add value to the story without distracting your audience:

- Break your data into small, digestible pieces. The more time your audience focuses on the data, the less energy they're going to put into listening to your story. Try the tips in these two books to create lightweight visualizations. Both books describe a process for stripping away needless information. You're trying to get to the bare minimum that you need to communicate something interesting. Turn off the text labels when you look at the visualization and see if the data still make sense. If you choose to display lots of data, then it is better to have a consistent flow of digestible chunks than a couple of charts that the audience needs to digest.

- Make a clear distinction between presenting data and storytelling. A great way to make this distinction is to use a clicker. Hold the clicker in your hand to change slides and present data. Put the clicker down and use both hands to tell your story. This gives your audience a chance to absorb the data so they don't have to listen to the story while processing a new visualization.

- Remember that data visualizations by themselves don't get you very far. If you want to tell a story about a great city like Chicago, don't just display a beautiful subway map. A wonderful data visualization like a subway map can tell you where to go, but it can't give you a reason to go there. Tell stories about the great food, the wonderful neighborhoods, and the sandy beaches of Lake Michigan. Those things spur your audience into action and might make them want to visit.

If you understand the limits of visualizations, you can get value from the added benefit. Just don't make the mistake of thinking that good visuals can take the place of an interesting story.

Eliminating Distractions

When telling a data science story, you're tempted to share a lot of data. Some teams feel it's a good opportunity to show off their work. Unfortunately, the Storytelling Session is not a good time to show the complexity of data science. As you've seen, a good Storytelling Session uses the data as a supporting actor and not the main character. That's one of the reasons you'll take anything out of your presentation that doesn't enhance the story. As you get closer to finalizing the content for your Storytelling Session, you should take more stuff out and put less stuff in.

There are two main places you want to be as neat and tidy as possible. First, as discussed in the previous section, make sure that your data visualizations are as clean as you can make them. Second, make sure that you have *just* enough characters, plot, and conflict to hold the story together. Strip away everything so you can get to the very essence of the story.

I once worked for an organization that had storytelling sessions that were driven by one of the directors. He started the session by congratulating the data science team for delivering great results. Then he talked about how important it was for the company to become more data-driven. He said it was a big part of the organizational strategy that was laid out by the stakeholders in the audience.

After ten minutes, the director started telling the meat of the story. Unfortunately, by this time, a lot of the audience had stopped paying attention. Some of the executives looked at their smart phones, while others just stared blankly at the first slide. This added information at the beginning of the presentation made it much more difficult for the director to engage the audience.

Now imagine if the director immediately engaged the audience. He says, "We think we found a way to better predict our customers' behavior. By looking at patterns, we're getting better at telling what our customers will buy before they even think of buying it." Then he tells a story about a typical customer, and even uses other characters in the story. With this approach, the audience is engaged from the moment he starts telling the story.

Remember that the audience wants to take something away from your story. As soon as you start your Storytelling Session, give them something interesting. As you continue to give them more, they will be drawn deeper and deeper into your story.

That's why you should work to strip away all the organizational norms that are usually at the start of every meeting. A storytelling session is a special occasion. Your research leads should never treat it like a typical status meeting. You don't have to congratulate the team. You don't have to show the importance of their work. When you give the audience what they want, they're more likely to attach meaning to your story.

You shouldn't need many data visualizations for your Storytelling Session. When you use them, try to eliminate any distracting information. Pull out any detailed data that can be easily summarized. Remove any text that you don't need from the image. When you look at the data visualization, ask yourself if there's anything that can be eliminated. If you remove it, would it affect what you are trying to communicate? Not every point needs a text label, and not every series needs to be spelled out. For example, if you have a time series that shows Monday through Friday, not every day needs a label.

When you eliminate distracting information, you're doing some of the audience's work for them. They don't have to think as much about the visualization and wonder what you're going to talk about. You can eliminate these obstacles between you and the audience.

Remember that less is more. A good Storytelling Session isn't like a fireworks show. You don't want to dazzle the audience with your sights, sounds, and swirls of color. It should be very simple and focused. There should be easy-to-read visualizations and a simple storyline with just a few memorable details.

Summary

In this chapter, you learned that a data science team should use stories to reveal the humanity behind the numbers. A great way to do this is using visualization with your storytelling. In addition, you found out that you should remove anything in your presentation that doesn't enhance the story. In Chapter 23, you'll find out how to use metaphors to ease in new ideas.

Using Metaphors

We live in a world filled with metaphors. We see them in common phrases. You can be as "busy as a bee," "heartbroken," or as "quiet as a mouse." They're common in literature. You've probably heard that metaphor that "all the world's a stage" or that Macbeth tried to look into the "seeds of time." Politicians use them in their speeches. There was President Reagan's "morning in America," and President Obama said the economy had run "into a ditch."

The reason you see so many metaphors is that they work. They connect something you know to something you don't know and make the unfamiliar seem more familiar. In their book *Metaphors We Live By*,[1] authors George Lakoff and Mark Johnson argue that metaphors are essential to how we think. We use metaphors to understand concepts such as love, war, and cooperation. They write that, "The people who get to impose their metaphors on the culture get to define what we consider to be true."

When you're telling a data science story, use metaphors as a way to ease in new ideas. Remember that metaphors make the unknown seem familiar, and when you hear a story about something familiar, you're more likely to connect it to some meaning.

---

[1]George Lakoff, and Mark Johnson, *Metaphors We Live By* (University of Chicago Press, 2008).

© Doug Rose 2016
D. Rose, *Data Science*, DOI 10.1007/978-1-4842-2253-9_23

From a literary perspective, a metaphor makes two things the same. Look at the phrase, "chain reaction," for example. You think about how a chain works with each link connected. When you think about something happening, that one thing impacts several other links. This is much easier to imagine than a term like "self-amplifying event."

In storytelling, just think of metaphors as anything that connects an unknown to something you know. That way, you don't have to worry too much about the nuances between metaphors, allegories, similes, and analogies. Just keep it simple. If you're equating two different things, think of that as a metaphor.

Data science involves a lot of difficult concepts, so there are already a few well-established metaphors: data warehouses, data mining, data lakes, technical debt, and panning for gold, just to name a few. This is the type of poetic language you want to use when describing difficult data science concepts.

Imagine that your team is working for a big movie studio. You want to figure out a way to use predictive analytics to decide how many screens on which to show a new movie. You don't want to show the movie on too many screens, because then you'll have a lot of empty seats in the theater. You also don't want to show it on too few screens, because people might not be able get tickets and skip the show entirely. Your data science team has gathered structured and unstructured data. You have a lot of structured data that shows people watched the movie trailer on many different web sites. You also have a lot of unstructured data that shows that there's a lot of movie buzz. People are talking about the movie a lot on social media sites such as Twitter and Facebook.

When your research lead tells the data story to the audience, instead of saying something like, "Our unstructured data analysis suggests that there's a lot of interest in this title," maybe she should say, "There is a lot of friendly chatter on social media sites that shows people really want to see this movie." That way, the audience immediately knows the value and source of this data. The audience also has an image in their head of how the data was created.

You can use metaphors in other ways as well. For example, "These are *hot tickets*" and "A few weeks after the release of the movie, we can expect a *cool-off period*." These metaphors make the story more interesting and fun, which keeps your audience engaged and helps them find some meaning in your story.

When you use metaphors, you're likely to break down the barrier between you and your audience. In data science, there is always some danger in using complex terms. You always risk creating a disconnect. A metaphor not only makes your story sound more interesting, but it also lowers the barrier to participate. Someone in your audience might be more likely to question the value of "friendly chatter" compared to "unstructured data analysis." The more your audience engages, the more likely they are to extract some meaning from the story.

Setting a Vision

In her book *Resonate: Present Visual Stories That Transform Audiences*,[2] Nancy Duarte covers techniques used to create a vision to spark change. One of the techniques she talks about is creating contrasts. This is the ability to separate the current context from a vision of the future. There are many times when you might want to use data science stories to create a new vision for your organization. For example, you see something new in the data and want to change direction to take advantage of the new insight. You may discover something in the data that makes a case for a new role in the organization.

Either way, setting a vision is one of the most challenging things to do when you're telling a story. A story that contrasts the present with the future requires a great deal of trust. No one is going to want to go on your new quest with you if they don't trust that you know the way. That's why one of the first things you want to do is help establish your own credibility. One way to do that is to use Engagement, Vision, and Authenticity (EVA). If you're a fan of the movie *Wall-E*, you should have no trouble remembering this approach ("Eeevva"). This technique is very valuable when you want to tell a story with a new vision for the future.

The first thing you want to do is engage your audience. Use some of the tricks we've covered in this book to draw them into your story. If you can't convince your audience that there's an interesting story to tell, they won't be motivated to make any changes. Here are some tips for you to remember:

- Focus on an interesting plot and have strong characters with memorable details.

- Set up a clear vision for the future. Help the audience picture the real changes in this vision of the future.

- Communicate this vision with authenticity. If it comes across as a sales pitch, your audience won't trust that your vision is in their best interest.

Creating an Interesting Plot

So let's say that you work for a power company. Your data science team comes up with a way to more efficiently distribute power. In your Storytelling Session, talk about how you currently distribute power. Explain that a lot of power gets wasted because it goes to areas where it's not needed. Suggest that you could use data science to create devices and route power in real

---

[2]Nancy Duarte, *Resonate: Present Visual Stories That Transform Audiences* (John Wiley & Sons, 2013).

time based on demand. Let them know that your team can predict how much power your customers will need based on a combination of unstructured and structured data. They can analyze data from the National Weather Service, and then compare that data to some of the unstructured data from social media. You could help tell the story using a common metaphor such as a "smart energy grid." You could also bring in case studies from other companies that are trying something similar, and then use critical thinking to compare your organization and highlight key differences. All of these ideas help create the plot of your story. You want the organization to go on a new quest, so make the characters interesting and establish some conflict between the present and the future.

Creating the Vision

The vision is the part of the story that has the greatest impact. Create a vision of the future where the organization uses data science to route power in real time and based on individual need. Talk about how this would be more environmentally friendly. For example, a wind farm in Texas may have a windy day, which would give excess power to a hot spell in Arizona.

You can also compare the system to a living organism. Use this metaphor to talk about how the system will adapt and breathe on a steady flow of data. All of these techniques will help engage your audience and solidify your vision.

Maintaining Authenticity

Finally, you want to maintain your authenticity. When you try the EVA approach, your audience needs to perceive you as someone who's internal to the organization. It's extremely difficult for consultants and people outside the organization to bring in the kind of authenticity you need to really engage your audience and establish a new vision. You need to come across as someone who cares about the organization and has a genuine interest in this quest.

Creating a vision of the future is one of the most challenging stories that you'll tell, but it's something you'll need to do if you want to make real changes. Try to remember the EVA approach as a way to make these stories more successful.

Motivating the Audience

The philosopher Plato once said, "Those who tell the stories rule society."[3] They do that because they motivate people to listen and make changes. You've seen how to build a story through different techniques that help engage your audience and communicate a meaning. Now it's time to bring it all together and get your audience to act. You'll want to take that level of engagement and turn it into something that motivates your audience to make changes.

There are seven steps that lead your audience into taking action. Each one builds on the other and ends with the new action item. You want your audience to leave your Storytelling Session motivated to try something new.

1. **Knowing your audience**: Figure out what motivates your audience. If you can pinpoint their needs, you can tailor the story to appeal to their fears or desires.

2. **Creating an emotional connection**: Use personal anecdotes and short vignettes to appeal to them on an emotional level.

3. **Providing context**: You can't talk about where you want to go without talking about where you've already been. Get your audience to understand why there's a need to do something new.

4. **Making your audience care about your characters and plot**: If the audience doesn't care about the organization or the data, they won't be motivated to act.

5. **Using metaphors to make the change seem more familiar**: You don't want your audience to be afraid of taking action. A good metaphor eases your audience into accepting something that feels new or dangerous.

6. **Using a clear contrast**: Use the context to set where the audience is and then use contrast to show where they need to go. This can be a new product or service, or a data story that shows that the organization should stop doing something.

7. **Creating a clear call to action**: If you've done a good job with the previous six steps, the audience will be ready to take some new action. State clearly what you want your audience to do differently. You don't want your audience to go back to doing the same thing they've always done.

---

[3]This quote has also been attributed to Aristotle, Hopi Proverb, Navajo Proverb, and Native American Proverb (unspecified).

Think about how you can use the seven-step process to tell a story about the running shoe web site. Imagine that the data science team has some strong evidence that customers are hesitant about buying shoes online. The data suggests that some customers are making frequent returns and then buying shoes elsewhere. The data science team asked some interesting questions and now the research lead wants to tell a story about what she found.

She starts out by creating a fictional customer who represents what they're seeing in the data. This customer buys everything online. With an online store, she finds it frustrating that she can't try on a pair of running shoes before she buys them.

So the research lead tells a story about how the company can expect an uptick in sales if they create some new storefronts in a few larger cities. The data science team uses some of its critical thinking skills to argue that many other organizations are experimenting with traditional storefronts. She creates the context that the web site is losing potential customers. She uses the metaphors "brick-and-mortar stores" and "virtual stores in the cloud." In her story, the research lead says that she wants to get the best of both types of stores. Then she talks about the plot in which the company creates these storefronts and connects the customer to this new experience. She follows the structure of the typical "quest" plot. She tells a story about how the organization is going someplace new. Then she creates a very clear contrast between the current web site and a beautiful new storefront. Finally, she ends the story with a call to action. She wants the audience to create a budget for this new venture.

She brings these elements together to motivate the audience to make changes. If you're the research lead, make sure that each of your Storytelling Sessions ends with a very clear and immediate call to action.

Summary

In this chapter, you learned how to engage your audience by focusing on an interesting plot and having strong characters and memorable details. This sets a clear vision for the future, helping the audience picture the real changes in this vision of the future. You also found out how to create a plot and a vision of the future. Finally, you found out that it's important for you to maintain your authenticity and come across as someone who genuinely cares about the organization. In Chapter 24, you will learn how to avoid storytelling pitfalls.

Avoiding Storytelling Pitfalls

One of the biggest challenges in storytelling is the idea that to be professional, you need to present data as a raw set of numbers. Many organizations believe that the data speaks for itself. The sheer power of the numbers will compel your audience to act. This is especially true if your culture is focused on objectives and compliance. In these organizations, you don't need to tell a story about how your project is on budget. You don't have to tell a story about the number of milestones you've completed.

When you're working in data science, you're trying to communicate something grander than simple status reports. You're trying to discover something new. Remember that the "science" in data science is about exploring data using the scientific method. This type of data is complex and needs to be interpreted. Your audience will look to your team to not only show them your well-designed reports, but also to help them understand the data's meaning.

Think about any time that you come in contact with complex data. Maybe you wanted weather information, or to see how a candidate was faring in an upcoming election. Both of these are complex data problems, which is why both are often inaccurate. The majority of people won't go deep into your

© Doug Rose 2016

D. Rose, *Data Science*, DOI 10.1007/978-1-4842-2253-9_24

data. Instead, they want to be told a story. They want to hear an interpretation of what *you* think about the data. Giving them too much data is not only unhelpful, but it might also be overwhelming.

Imagine that you're watching a political show and the commentator puts up four bar charts. He says, "As you can see, the data speaks for itself." Most people would just change the channel without looking at the reports. It's the same with data science. If your story is just using charts, then your audience will quickly dismiss your presentation. A good data story uses reports as a garnish to enhance a larger dish.

There are a couple things to watch for to make sure that you don't over-rely on your data visualizations.

- Examine your presentation. If you're using slides, how many do you have? If it's an hour-long presentation and you have thirty slides, then you're not telling a story. You're probably just presenting the data.

- Examine how much time it's taking you to prepare your data. It's great to make sure that your charts are clear. Just remember that the charts are one of the first things that your audience forgets. If you want to have maximum impact, focus on the things your audience will remember. Your audience is more likely to remember a clear, interesting story.

If your organization has a very conservative management culture, it might be very difficult to tell stories. In these organizations, it is often politically safer to present the charts and leave it to the managers to interpret the data. You might be tempted to just portray yourself as an impartial presenter. The problem with this approach is that if you're on a data science team, you're still responsible for the outcome. So you'll be on the hook for whatever they decide is the best interpretation of your data. In these situations, it's usually still a better strategy to express your opinion through a well-told story. That way, at least you have some control over whatever happens with your results.

Finally, it's often very difficult for new teams to accept that you can create a story from data. Some data just looks like lifeless columns of decimal numbers. It's a real challenge for those teams to look at those digits and reverse engineer the activity that created them. Frankly, it's one of the biggest challenges of being on a data science team. The best way to avoid this is to humanize your reports. Don't call a report "Upcoming consumer trends." Instead, call it something like "What people are buying." These little steps can make it easier to think about your data as reflecting real-world events.

Like any skill, data storytelling takes time to improve. Start thinking about the key characteristics of a story, such as plot and conflict. Then work to present your data in an interesting way. Over time, your stories will become more robust and interesting. You might even make stronger conclusions and braver interpretations. Try to remember to have fun with your stories and your audience. It will improve your stories and make you a more interesting storyteller.

Summary

In this chapter, you saw how to avoid some common pitfalls when telling data science stories. Be sure to look at your presentation for warning signs that you are putting too much emphasis on just the data. Also, watch out for organizational culture challenges that might make it more difficult to tell interesting stories. In Chapter 25, you'll see how to put the five parts of this book together and make some real changes in your organization.

Finishing Up

Well, that was fun. We covered a lot of territory. First, we started out with a foundation for understanding data science. Then you found out how to create data science teams. You also learned how to map your data science team roles to people already in your organization. Then you looked at a new data science life cycle (DSLC) framework for working as a team. You explored delivering short sprints of value over time, which allows you to pivot your work to adapt to feedback and create better insights. Next, you learned how to reason with your data and use strong-sense critical thinking. Finally, you learned how to tell a compelling data science story. Good storytelling is the bridge between what you learn and what you can tell others. Without that bridge, you won't get the useful feedback you need to connect your team's insights to real business value.

So, what's next? There is one last challenge. You need to understand your organization's culture so you can help them make the change.

Starting an Organizational Change

Organizational change management is a well-explored field. You're already armed with the knowledge you need to start data science in your organization, but in this chapter, you'll learn some of the other tools you can use to help change the mindset of your organization.

I started out this book by describing my job in the early 1990s working at Northwestern University's Academic Computing and Networking Services (ACNS). At that time, the office was located in downtown Chicago. Every morning, I took the elevated train from the north side red line to the downtown stop at State and Chicago Avenue. Each time I walked through these north side neighborhoods, I would pass by one house with a small fenced-in yard. Behind the fence there was a dog who would scratch, claw, and bark every morning as I made my way to the station. It was our routine. Something that the dog and I shared each morning. I would walk to the trains, and then the dog would scratch and claw and bark. Sometimes I would even see its shiny brown head pop over the top of the fence.

One morning the dog was unusually motivated. The fence planks rumbled like a loud speaker as the dog threw its body against the creaky wood. I glanced

© Doug Rose 2016
D. Rose, *Data Science*, DOI 10.1007/978-1-4842-2253-9_25

at the fence and then went back to my steaming NWU coffee tumbler. The dog managed to get one of its limbs over the side. It then used this leverage to roll on its back and with super strength flipped over the side of the fence. With some lost grace the dog stood up and seemed as surprised as I was at its accomplishment. It looked back at the fence with a hint of regret, and then at me. Our eyes locked and, in that instant, we realized that something had changed. Neither of us knew what to do.

I think about this story when I see organizations making big changes. Often, all the passion and the effort goes into scratching and clawing to try something new. In reality, that's not how organizations change. Large organizations aren't usually moved by passion. They're moved by relentless pursuit of long-term, practical improvements.

Most organizations fail to make big changes for three reasons:

- They don't understand the change. Organizations often don't understand the value of a new data science mindset. The key players don't have a clear sense of what an exploratory and empirical organization would look like. They don't have a clear picture of what their organization would look like at the end of the journey. So they might have a few teams who try a new data science mindset. These teams will try new things, but they won't have a plan for where they're going. No one has communicated the benefit of making an organization change.

- They don't have a good sense of their own culture. They don't have a vision of how a data science mindset will fit into their larger organizational norms. They haven't considered whether their organization would accept the change. There might be a big disconnect between a data science mindset and the way your organization operates. Before you implement the data science change, you need to have an objective idea of your larger organizational culture.

- They don't have a real plan for changing. The organization doesn't know the practical steps they need to take to make the change. They might understand the data science mindset and feel like they have the right culture, but they don't know how to connect the two. This can be a big challenge because in most organizations, thinking about data using a scientific method is a big change. If you don't treat it as an organizational change, you'll quickly run into insurmountable challenges. Even changes that are popular and widely accepted won't always fit your organization's culture. Well-accepted practices won't necessarily work well for your teams. If you don't treat a new data science mindset like an organizational change, you're in real danger of failure. You might have a few pockets of innovation, but it will be challenging to make lasting changes.

In this chapter, you'll learn about the different types of organizational cultures and how to identify *your* organization's culture. Then you'll be introduced to a resource that will help you learn how to get your organization over the fear of making the change.

Understanding Organizational Culture

For most organizations, the first step to implementing change is to better understand the organization's culture. Organizational culture is basically the stuff that people do without even thinking about it. Former MIT professor Edgar Schein wrote a terrific book[1] on the subject. He came up with a more sophisticated definition. He says that an organization's culture is

"A pattern of shared basic assumptions that the group learned as it solved its problems that has worked well enough to be considered valid and is passed on to new members as the correct way to perceive, think, and feel in relation to those problems."

One of his key points is that organizational culture is deeply ingrained. These are the things that people do without asking. They are the assumptions that a group learns and teaches its new members. That's what makes an organization's culture incredibly difficult to change. In an organization, people have a success pattern. New people are taught these things and told this is the right way to do things when they're hired into the organization. Therefore, this "culture" is seen as the correct way of working and is very difficult to change. The business analysts and project management officers have all accepted that their organization's culture is the way to get things done. When someone in that culture wants to turn things around, it's often seen as backwards and incorrect. This is especially true when you're talking about a new mindset.

That's why so much of the focus goes into scratching and clawing, and trying to make the new change. The only challenge with this strategy is knowing what to do when you start to implement some of these changes. What happens when you actually get some of the people to flip over the wall? In many organizations, that conversation never takes place. All of the effort goes into making the change. But what happens after you make the change and some of the people don't easily accept the new mindset? Before you put all of your effort into scratching and clawing, you need to evaluate the culture at your organization.

---

[1] Edgar H. Schein, *Organizational Culture and Leadership,* Vol. 2. (John Wiley & Sons, 2010).

Thankfully, there's a pretty good resource out there for identifying your organization's culture. It's a book called *The Reengineering Alternative* by William Schneider.[2] He created four categories to help you identify your culture. Each one focuses on something different (Figure 25-1). The four types are:

- Control
- Collaboration
- Cultivation
- Competence

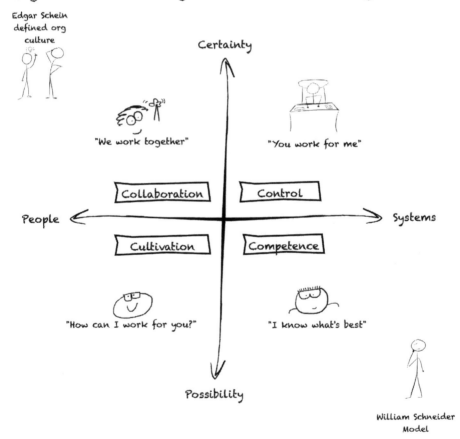

Org Culture = The things we do without thinking about it

Edgar Schein defined org culture

Certainty

"We work together"

"You work for me"

Collaboration   Control

People   Systems

Cultivation   Competence

"How can I work for you?"   "I know what's best"

Possibility

William Schneider Model

Figure 25-1. Customers who live near a gym

[2]William E. Schneider, *The Reengineering Alternative: A Plan for Making Your Current Culture Work* (Richard D Irwin Publishing, 1994).

In his book, he offers a questionnaire that you can pass around your organization. The questionnaire helps you determine which one of these categories best describes your organization. There may be some overlap in the categories. You may be a company that puts a lot of value on competence, but you might also put a fair amount of value on collaboration. It's not that each one of these categories will encompass your entire organization. Still, you'll probably see one type of organizational culture that clearly stands out.

Each one of these cultures has its own strengths and weaknesses. One culture might readily accept new changes, while another might fight even the most sensible change. The key is that once you understand your organization's culture, you will have an easier time determining how much of the data science mindset will expand beyond your team.

Control Culture

Let's start with the **control culture**. This culture has a tendency to be very authoritarian. Think of it like a wolf pack. Such companies tend to have a conservative management style and put a lot of emphasis on hierarchy. Everyone in a control culture knows who they work for and who works for them. In a control culture, there's a lot of emphasis on compliance. The role of the individual is to comply with the requirements of their supervisor. The head of these organizations communicates a vision, and then everyone who works for them is responsible for implementing this vision.

There are also people in the organization who ensure everyone complies with the vision. A control culture prefers employees to stay within their functional areas. Individuals don't usually move around very much. A lot of the authority in a control culture comes from roles and titles. Directors have authority over managers, and managers have authority over supervisors. The title communicates the level of authority.

Because there's so much emphasis on compliance, the decision-making in a control culture has a tendency to be very thorough. There is a push for certainty at the highest levels. The C-suite or the directors want to know when a decision has been made. They want someone who works for them to "sign off" on the decision. The way that you make big changes in these organizations is to get someone at a high level, such as a chief data officer (CDO) or a chief information officer (CIO), to "sponsor" the change. Without this sponsor, it will be very difficult to make any lasting changes beyond your team. Unfortunately, control cultures put so much emphasis on certainty that it's extremely difficult to get high-level sponsors to agree to make big changes. Big changes are almost inherently risky. The most common way that control cultures make big, risky changes is when they have little alternative. These organizations tend to be more conservative and the norms in the organization favor order and certainty. They usually like predictable processes. That's why a lot of these organizations will gravitate toward changes where there's a lot of reliance on a large system. These systems are set up so that everybody knows their part as well as their place in the larger process.

Data science teams often have difficulty in control cultures because data science team roles are more flexible than roles in a control culture. Also, exploration is by definition uncertain. Data science teams in a strong control culture don't usually have easy access to the data or the authority to make decisions. Control cultures favor functional areas in organizations and strong departmental barriers. That makes it difficult for project managers on these teams to break through data silos. Even with these challenges, many data science teams still work in a strong control culture. Control cultures are very prevalent in large organizations, which are the same organizations that are likely to have massive amounts of interesting data. In many ways, organizations with strong control cultures often benefit the most from a well-functioning data science team.

Competence Culture

A second very common culture is the **competence culture**. This culture is prevalent in software development organizations. A typical competence culture is one in which a group of software developers created a tool that became very popular, and the developers became the de facto managers. This competence culture was set up as part of their organization. The leadership focus for competence culture is about setting the standards and creating tasks. They distribute these tasks based on each employee's level of competence. So the management style is very task driven. The management team tries to be analytical in how they distribute their tasks. It's about who will do the best job finishing the work. Organizations with a strong competence culture have a tendency to be a matrix organization. In a matrix organization, an employee might have several managers. You might have a quality assurance developer reporting to a quality assurance manager and also reporting to a software development manager. This means that you have a lot of employees who want to be specialists and much of their focus will be on specializing. You don't want to be a quality assurance developer who knows too much about development. Then your software development engineer might give you tasks and you'll be quickly overwhelmed. So there's a lot of emphasis on specializing.

In a competence culture, a lot of the power and authority comes from developing a high level of expertise. The decision-making in these organizations has a tendency to be very analytical. Such companies break down a problem into components and then distribute those components to different employees. They view the organization as an engineering problem. Often, when engineers have problems, they break them down into component parts. These organizational cultures manage change by driving big goals. They take a big goal and break it down into tasks in order to move forward. They have a tendency to be very professional and in these competence cultures, with a strong sense of meritocracy. You could come in as a low-level employee or an intern, and if you specialize and develop a high level of expertise, you move your way up through the organization.

Organizations with a strong competence culture also have a tendency to have a very intense pace. They're not always the easiest places to work. Competence cultures can also have a difficult time embracing a data science mindset. Data science tends to be interdisciplinary. You have to know a little bit about statistics, mathematics, storytelling, and programming. Cultures that put a lot of emphasis on specializing might not have an easy time with this approach. You don't want your data analyst to refuse to help out when you need to tell a good story. You also don't want your project managers and research leads to feel unqualified to ask good questions. Competence cultures place a lot of emphasis on being an expert. That's why there's a high bar for expertise. That can be a challenge because in a data science team, it's often outsiders that will give you some of your best questions.

Cultivation Culture

The third type of culture is the **cultivation culture**. This is the rarest of all four types. In this people-driven culture, the leader focuses on empowering and enabling people to become the best possible employees. The managers like to make sure everyone is happy. They want them to enjoy being part of the organization and there's a lot of emphasis on employee surveys. These organizations have a tendency to be set up with a wheel of authority, with the employee in the center and all their resources around them. Each one of their managers is like a spoke in the wheel. The people around them try to help the employee figure out what they need to be their best.

In a cultivation culture, there's a lot of emphasis on expressing yourself. The managers focus on developing and growing the employees. They want to build everyone up. The leadership is typically focused on being charismatic. If you're a charismatic person in a cultivation culture, you can quickly become an authority—even if you just started at the company in a low-level position.

Managers focus on cultivating the strengths of other people. A cultivation leader rises in the organization by solving problems through the talent of their team.

In a cultivation culture, a lot of value is placed on being a generalist. You don't want to knock on someone's door and hear that they can't solve your problem.

One thing that you never see in a cultivation culture is someone who's stuck in the system. You'll see much less emphasis on departments and processes in the cultivation culture. In addition, decision-making in these organizations can be difficult because it is highly participatory and organic. Everyone on the team wants to drive toward consensus.

Millennials and people who are under the age of 30 have a tendency to be successful in these cultivation cultures.[3] A lot of younger employees are especially driven toward finding consensus. Organizations that are run by young entrepreneurs tend to put a lot of value in this cultivation culture, and are more likely to embrace change and adapt to new ideas. They assume that change is part of the cultivation process. They have participatory meetings where people talk about change. Then, once they decide that a change is good for their company, the change is embraced quickly. Growth and development is encouraged in these organizations, and they have a tendency to be pretty fun to work at because people are free to make mistakes.

However, the challenge with these cultivation cultures is that they tend to move slowly in the decision-making process. As you can imagine, it takes a long time for big groups to come up with a decision everyone agrees on. True cultivation cultures are rare. Some organizations may feel like they have a cultivation culture, but if you look closely, you'll see that they don't really follow a lot of the key practices. A lot of these organizations are just control cultures with a thin veneer of a cultivation culture.

Collaboration Culture

The fourth and final culture is the **collaboration culture**. This is almost as rare as the cultivation culture. You really don't see this much in information technology because it just doesn't play to the leadership style. This type of culture occurs more in training organizations. The leaders in a collaboration culture tend to be team builders and coaches. Their management style is very democratic, but it's not quite as ad hoc as the cultivation culture. There's less need to get everybody on board, but you'll still have a group of managers who work closely together to come up with interesting ideas. That's the whole point of the collaboration. Such companies have a tendency to have group clusters instead of a top-down hierarchy like you see in control cultures. They still put a lot of emphasis on being a generalist.

The big difference between the collaboration and cultivation cultures is that, in the case of the former, the authority comes from relationships. Sometimes you'll see this in family-run businesses. The closer you are to people at the head of the organization, the more authority you have. The top people collaborate more closely. They have a tendency to make decisions with brainstorming meetings and some experimentation. They're a little more open to

---

[3]Brad Karsh and Courtney Templin, *Manager 3.0: A Millennial's Guide to Rewriting the Rules of Management.* AMACOM Div American Mgmt Assn, 2013.

change than the control or competence cultures. This helps if the organization is trying to embrace a data science mindset. If you have a collaborative culture, it's not difficult for your organization to accept the change. However, there are some key components of data science that a collaboration culture might find pretty difficult. An effective data science team has to have the authority to pursue new ideas and make mistakes. That authority is pushed down to the team level. Collaboration cultures still have a tendency to have the authority at a high level. They're just a little bit more democratic than a control culture.

Identifying Your Culture

Now that you've seen these four different types of cultures, you need to figure out which one most closely matches your organization (if you haven't already). Remember that a collaboration culture or cultivation culture has an easier time embracing the key components of the data science mindset. Individuals in these cultures also change more easily because they already have a natural tendency to be generalists in their organization.

If you have a strong control culture, your organization gravitates toward the traditional big systems approach. These organizations tend to create large processes in which everyone knows their part. In many ways, that is the opposite of the small, self-organized teams you want running data science projects. These cultures also often have trouble embracing some of the key components of the scientific method. Experimentation and exploration are inherently unpredictable. These control cultures usually favor complex processes with predictable outcomes. If you have a very strong control culture, you may want to start out by having a few discrete teams. If these teams are successful, you might have the opportunity to make some larger organizational changes. Keep in mind that these cultures often have the most difficulty changing.

If you have a competence culture, your organization might have similar challenges trying to embrace a data science mindset. Much of the scientific method is breaking things down into interesting questions so the team can pan through these questions to look for larger insights. This is much different from how a competence culture usually views work. They see a large problem as something that can be broken down into tasks. It's very analytical, but not empirical. A team in a competence culture needs to know exactly what they'll be doing to finish the work. These organizations also have a tendency to have highly specialized teams. Each person in this team is supposed to have his or her own expertise. In many ways, this makes the team less collaborative. Each person is an authority for their own area. Again, this makes it more difficult to embrace a data science mindset. You want your data science team to be able to ask interesting questions. These questions might come from the research lead, but they might also come from the data analysts or even the project manager. Everyone on the team is assumed to be competent enough to ask interesting questions. That can be a bit of a change for a strong competence culture.

Making the Change

After you've identified your culture, you can start to make changes. One of the best books on organizational change is *Fearless Change*.[4] This book identifies patterns of how organizations accept or reject changes. You can take what you've learned about your culture and use it to identify which patterns will work best in your organization. Even in a control culture, you can sometimes set the stage for later changes. This book is geared toward "powerless leaders." These are leaders who don't have any implicit or explicit authority in the organization. So you don't have to be a CEO or director to try to implement change in your organization.

This book is perfect for leading a change in your organization's mindset. Even in a strong control culture, you don't necessarily have to be a director or manager in the C suite to start the change process. You just have to be someone with a good idea who wants to introduce the change. The book has 48 patterns for change leaders. You can mix and match these patterns to come up with an overall strategy based on the culture of your organization.

One of the most helpful guides in the book is the "myths" of organizational changes. Several of these myths might block your efforts to make changes. The most common are, "If it's a good idea, it'll be easy to convince others to accept it," and "All you need to implement a new idea is knowledge and an effective plan." Another is the idea that if someone is skeptical about a change, you can overrun or ignore them. The book gives you strategies for how to deal with skeptics and how to listen to them. You shouldn't ignore your skeptics; they might be right and see something that you don't.

Another myth that the book points out is that you can be a change agent in your organization and work alone—that just the power of your knowledge, charm, and PowerPoint will be enough to make a change. What this book tries to show is that you need to create groups of people who can help you make changes. If you're powerless, the best thing you can do is create consensus with a small group of people and then get the change moving forward.

---

[4]Linda Rising and Mary Lynn Manns, *Fearless Change: Patterns for Introducing New Ideas* (Pearson Education, 2004).

The final myth is the misconception that if you convince someone of the change, they'll stay convinced. One of the things that the book does well is represent the change almost as a juggling act, in which you convince someone that the change is worthwhile, but you still have to go back to them every now and again. In that case, ask them if they are still on board and if they still support the change. Even if people accept the change and accept the idea, they might slip back into their old ways. This is particularly true of control organizations when you're trying to make big changes.

Fearless change is based on the notion that people accept ideas at different rates. These people naturally fall into one of several groups:

- **Natural innovators**: When this group sees something new, they are the first to accept the change. You'll see this with your data science mindset. Some innovators in your organization will be very interested in using the scientific method to try to better understand your data.

- **Early adopters**: This group is interested, but they want to hear more. They might think it's a good idea but are not quite as motivated as the innovators.

- **Early majority**: This group is the biggest chunk. They are the people who think the idea is interesting but wait to see what other people say before they get on board.

- **Late majority**: This group says, "Okay, if everybody's on board, I'll get on board, but I don't really want to be the first one."

- **Laggards**: These are the people who say, "I really like the way things are done now, and I don't see why we need to change."

This idea that people accept change at different rates is helpful when you're thinking about a large-scale organizational change. You're trying to shift to a data science mindset, and this is usually a big change to the culture. That's why you often want to make sure that you get the early majority people on board. You can use your innovators as a way to recruit them and try to get enough consensus so that you can push your agile transformation and gain a little bit of momentum.

Another thing to remember is that when you're trying to change the culture in an organization, you want to speak not only to people's heads, but also to their hearts. You want to be able to talk to them on a level where they make an emotional connection with the change. This allows you to motivate your innovators and encourage your early adopters. You don't just want to talk about productivity. Talk about your data science mindset as a way to better understand your customer. Maybe even approach it as a way to make an emotional connection to your customer. The innovators and early adopters are usually looking for something interesting to get behind. If you can stimulate their interest, they will usually stick with you for the entire organizational change.

The book also goes through the idea that you can't overrun people. You can't ignore the cynics and skeptics in your organization. Oftentimes, the cynics and skeptics are right and have good points. I'll often see organizations in which the change leaders try to ignore these people and perceive them as impediments. They'll dismiss them as laggards who don't want to make big changes. If you ignore the cynics and skeptics, chances are you're going to run into more problems. These people are often the first ones to point out challenges. You should listen to them closely and understand what they're trying to say. Try to convince them that this transformation can take place, even though they are cynical or skeptical about a few pieces. Your skeptics will want to weigh the benefit of your change against the cost of the effort.

Summary

In this chapter, you've learned about the different types of organizational cultures and how to identify your organization's culture. You were then introduced to a resource that will help you learn how to get your organization over the fear of making changes. In the next section, I'll give you some parting thoughts as the grand finale.

The Grand Finale

So this is it. We're at the very end. I hope you've enjoyed this book on data science and you've a better idea of how to form teams that can ask interesting questions and deliver real business value. One of the key points I've tried to make is that data science is more than just a set of practices. It's about having an exploratory and empirical mindset. There are many books on the tactical aspects of data science. What I've tried to do here is different. I've tried to show that tactics have a much shorter lifespan than a shift to a larger data science mindset. If you learn R, Python, statistics, or Hadoop, then you'll have some of the tools you'll need, but using the tools alone won't make you a data science team. Remember to focus on the "science" in data science. This data science mindset will give you the freedom to use these tools in much more interesting ways. *Thinking* about your data in new ways is far more challenging than just downloading new tools and software. A new data science mindset is a real challenge, but it will be a more rewarding and productive way to work with your data.

I hope you enjoyed this book and have fun asking great questions, gathering up insights, and learning more from your data.

Index

© Doug Rose 2016
D. Rose, *Data Science*, DOI 10.1007/978-1-4842-2253-9

Get the eBook for only $4.99!

Why limit yourself?

Now you can take the weightless companion with you wherever you go and access your content on your PC, phone, tablet, or reader.

Since you've purchased this print book, we are happy to offer you the eBook for just $4.99.

Convenient and fully searchable, the PDF version enables you to easily find and copy code—or perform examples by quickly toggling between instructions and applications.

To learn more, go to http://www.apress.com/us/shop/companion or contact support@apress.com.

All Apress eBooks are subject to copyright. All rights are reserved by the Publisher, whether the whole or part of the material is concerned, specifically the rights of translation, reprinting, reuse of illustrations, recitation, broadcasting, reproduction on microfilms or in any other physical way, and transmission or information storage and retrieval, electronic adaptation, computer software, or by similar or dissimilar methodology now known or hereafter developed. Exempted from this legal reservation are brief excerpts in connection with reviews or scholarly analysis or material supplied specifically for the purpose of being entered and executed on a computer system, for exclusive use by the purchaser of the work. Duplication of this publication or parts thereof is permitted only under the provisions of the Copyright Law of the Publisher's location, in its current version, and permission for use must always be obtained from Springer. Permissions for use may be obtained through RightsLink at the Copyright Clearance Center. Violations are liable to prosecution under the respective Copyright Law.

54109096R00155

Made in the USA
San Bernardino, CA
07 October 2017